Developing a Plan for the Planet

This book is dedicated to the millions of people who live in poverty, without access to education, clean water, health facilities or employment … and who have no voice of their own.

Developing a Plan for the Planet

A Business Plan for Sustainable Living

IAN CHAMBERS
and
JOHN HUMBLE

GOWER

© Ian Chambers and John Humble 2011

All rights reserved. No part of this publication may be reproduced, stored in a retrieval system or transmitted in any form or by any means, electronic, mechanical, photocopying, recording or otherwise without the prior permission of the publisher.

Published by
Gower Publishing Limited
Wey Court East
Union Road
Farnham
Surrey
GU9 7PT
England

Gower Publishing Company
Suite 420
101 Cherry Street
Burlington
VT 05401-4405
USA

www.gowerpublishing.com

Ian Chambers and John Humble have asserted their moral right under the Copyright, Designs and Patents Act, 1988, to be identified as the editors of this work.

British Library Cataloguing in Publication Data
Chambers, Ian G.
  Developing a plan for the planet : a business plan for
sustainable living. -- (Gower green economics and
sustainable growth series)
  1. Sustainable development. 2. Environmental economics.
  3. Social responsibility of business.
  I. Title II. Series III. Humble, John William.
  338.9'27-dc22

  ISBN: 978-0-566-08911-4 (hbk)
      978-1-4094-0682-2 (ebk)

Library of Congress Cataloging-in-Publication Data
Chambers, Ian, 1939-
  Developing a plan for the planet : a business plan for sustainable living / by Ian Chambers and John Humble.
       p. cm. -- (Gower green economics and sustainable growth series)
  Includes bibliographical references and index.
  ISBN 978-0-566-08911-4 (hardback) -- ISBN 978-1-4094-0682-2 (ebook)
  1. Sustainable development. 2. Renewable energy sources. 3. Sustainable living. I. Humble, John. II. Title.
  HC79.E5C44 2010
  658.4'083--dc22

2010000928

Printed and bound in Great Britain by the
MPG Books Group, UK

'If a visitor were to arrive from outer space and looked at planet earth, their immediate request would be to ask to see the manager.'
William S. Burroughs

'Human progress is neither automatic nor inevitable.
We are faced now with the fact that tomorrow is today. We are confronted with the fierce urgency of now. In this unfolding conundrum of life and history there is such a thing as being too late.
We may cry out desperately for time to pause in her passage, but time is deaf to every plea and rushes on.
Over the bleached bones and jumbled residues of numerous civilisations are written the pathetic words: Too late.'
Martin Luther King Jr.

'It is not the strongest of the species that survives… nor the most intelligent that survive. It is the one that is most adaptable to change.'
Charles Darwin.

Crisis = Danger + Opportunity[1]

---

[1] As reported in the Nobel Lecture by Al Gore on the American Rhetoric Website: www.americanrhetoric.com/speeches/algorenobellecture.htm – 34k.

The Green Economics and Sustainable Growth Series

Series Editor: Miriam Kennet

The series provides business decision makers, policy makers and academics with realistic and dispassionate guides to the implications of green economics; embracing the four pillars of green economics: Policy; Academia; Business; and Civil Society, the series presents cutting edge applied research and draws on the experience of businesses around the world. Here is a credible, detailed and commercially realistic agenda to engage leading businesses, regulators and policy makers, facing uncertainty and transformation in environmental, social and financial systems bearing on the future competitiveness and even the survival of business operations.

About the Editor

Miriam Kennet is the Founder and Editor of the *International Journal of Green Economics,* which is published by Inderscience and can be seen on www.inderscience.com/ijge, and on the Editorial Review Board of several other journals including *International Journal of Ecological Economics and Statistics*, and the *International Journal of Industrial Ecology*.

Miriam is Co-founder and Director of the Green Economics Institute.

Titles currently in the Series include:

Developing a Plan for the Planet
A Business Plan for Sustainable Living
Ian Chambers and John Humble

The Economics of Abundance
A Political Economy of Freedom, Equity, and Sustainability
Wolfgang Hoeschele

Contents

List of Figures	xxi
List of Tables	xxv
About the Authors	xix
Acknowledgements	xxi

1		**Why We Need a Plan for the Planet**	**1**
		The Purpose of This Book	6
PART I		**UNDERSTANDING OUR CURRENT SITUATION**	**13**
		From Flint Chip to Silicon Chip	13
		However, There is a Problem …	14
		The Perfect Storm?	15
		Crisis Can Also Create Opportunity	16
		A Word on Complacency, Fear and the Need for Cautious Optimism and Action	17
		Easter Island Case Study – The Collapse of a Civilisation	18
		Why Take a Business-planning Approach?	21
		A Word on Idealism	22
PART II		**UNDERSTANDING THE KEY GLOBAL CHALLENGES**	**25**
		Understanding the Issues	25
		Executive Briefs	27
2		**Executive Brief No. 1: Population Growth**	**29**
		The Current Situation	30
		Opportunities and Best Practice	34
		The Role of Government, Business and People	36
3		**Executive Brief No. 2: Climate Change**	**39**
		The Current Situation	40
		Opportunities and Best Practice	43
		The Role of Government, Business and People	47
4		**Executive Brief No. 3: Energy Supplies**	**57**
		The Current Situation	58
		Opportunities and Best Practice	60
		The Role of Government, Business and People	62
5		**Executive Brief No. 4: Water and Food Supplies**	**69**
		The Current Situation – Water Supplies	70

	Solutions and Best Practice: Water Supplies	74
	The Role of Government, Business and People	76
	The Current Situation – Food Supplies	78
	Solutions and Best Practice – Food Supplies: The Role of Government, Business and People	82
6	**Executive Brief No. 5: Planet Sustainability and Biodiversity**	**89**
	The Current Situation	90
	Opportunites and Best Practice	95
	The Role of Government, Business and People	96
7	**Executive Brief No. 6: Extreme Poverty**	**101**
	The Current Situation	102
	Solutions and Best Practice	103
	Opportunities	104
	The Role of Government, Business and People	105
8	**Executive Brief No. 7: Global Health**	**109**
	The Current Situation	110
	Opportunities and Best Practice	113
	The Role of Government, Business and People	115
9	**Executive Brief No. 8: Universal Education**	**119**
	The Current Situation	120
	Opportunities and Best Practice	123
	The Role of Government, Business and People	123
10	**Executive Brief No. 9: Conflict and Peace**	**127**
	The Current Situation	128
	Opportunities and Best Practice	133
	The Role of Government, Business and People	134
11	**Executive Brief No 10: Financing a Sustainable World**	**139**
	The Current Situation	140
	Solutions and Best Practice	140
	The Role of Government, Business and People	143
	How Much Will it Cost to Address These Strategies?	145
	Putting the Finances of Sustainability in Perspective	147
	The Problem of Interconnectivity	149
12	**Executive Brief No. 11: The Challenge of Interconnectivy – The Perfect Storm or the Perfect Opportunity?**	**149**
	The Perfect Storm?	153
	The Perfect Opportunity?	154
	The Key Next Steps: Developing a Plan for the Planet	155

PART III	**DEVELOPING A PLAN FOR THE PLANET**	**157**
13	**Building a Global Vision for Planet Earth**	**159**
	Why a Vision for Planet Earth?	159
	Building a Vision for Our Plan for the Planet	160
	Developing a '2030 Mission'	166
14	**Global Objectives and Strategies: Addressing Our Key Global Challenges**	**167**
	Involvement of the Global Community	169
	Objectives, Strategies and Action Plan Summary	169
	Business Plan Summary	171
	Organisational and Personal Action Planning: What, When, and How?	183
15	**Taking Responsibility: Translating Understanding into Action**	**183**
	Developing Your Plan for the Planet (Government, Business or Personal)	185
	Best Practice Global Management	189
PART IV	**MANAGING A PLAN FOR THE PLANET**	**189**
16	**Ten Global Management Best Practices**	**191**
	1. Effective Business Planning	191
	2. A Customer-driven Organisation	193
	3. Sensitivity to the Outside Environment	195
	4. Thinking Global/Acting Local – Global Partnerships and Teaming	196
	5. Understanding Competition	198
	6. The 80:20 Rule – Focus	200
	7. Productivity and Continuous Improvement	200
	8. Mastery of Information Technology	201
	9. Shared Values	205
	10. Effective Change Management	206
17	**Global Management Best Practices: Applications to Managing a Plan for the Planet**	**211**
	Application of Global Management Best Practices to Public Sector Management	211
	Achieving Best Practice	212
	Application of the Best Practice Global Management to International Organisations	213
18	**Global Management Best Practices: A Health Check**	**215**
	Application of the Ten Best Practice Global Management Principles	216
	Building a Sustainable World – An Effective Manager's Health Check	216
	A. Organisation Structure and Roles	217

	B. Organisation Dynamics	219
	C. Management Information	221
	D. Managing Change	223
	E. Managing Innovation	224
	F. Managing Motivation: Leadership Responsibility and Effective Teaming	226

PART V	DELIVERING A PLAN FOR THE PLANET	229
19	**Leveraging the Triangle of Change**	231
20	**Leveraging International Organisations and Government**	233
	Achieving Best Practice Global Management in International Organisations	235
	Opportunities for Increasing Management Effectiveness	242
	National Government	244
	Opportunities for State Owned Enterprises	247
21	**Leveraging the Business Contribution**	247
	The Business Contribution in Context	248
	Opportunities for Multinational Companies	252
	Social Contribution: Risks and Benefits	256
	Corporate Social Reponsibility Audit	258
	The Opportunity of the New Philanthropy	264
	Opportunities Presented by the Social Challenges of Supply Chain Management	266
	Conflict Management: Does Business Help or Hinder?	269
	Corruption: A Global Challenge for Business	270
	Investing to Fight the Ten Global Challenges	274
	Expanding Partnerships	275
	Business in the 21st Century: Corporate Global Citizens	278
	Building Bridges: Working Together	281
22	**Leveraging the Power of the People**	281
23	**Embracing the New Green Revolution**	285
	Faith Driven Power	287
24	**Embracing the Spiritual Imperative**	287
	Buddhism	288
	Christianity	288
	Hinduism	288
	Islam	288
	The Practical Power To Influence Opinion	289
	Working Together	289

| 25 | **Delivering Our Plan for the Planet** | **291** |

Epilogue   *301*
Appendix 1: Millenium Development Goals   *305*
Appendix 2: Useful Websites   *309*
Bibliography   *311*
Index   *315*

List of Figures

Figure 2	Road map to a sustainable world	8
Figure 2	Road map to a sustainable world concluded	9
Figure 3	Save Planet Earth – for all of us! It's the only home we've got!	12
Figure 4	From flint chip to silicon chip!	13
Figure 5	The perfect storm?	14
Figure 6	A clear relationship between population growth, greenhouse gases and climate change	15
Figure 7	Easter Island	18
Figure 8	I wonder whether he realises that he is cutting down the last tree?	19
Figure 9	A good case of bad planning?	21
Figure 10	What on earth are we doing to our Planet?	26
Figure 11	How many people can Planet Earth sustain?	29
Figure 12	Forecast world population growth (medium scenario)	30
Figure 13	I think I know where we might need to place a little focus!	31
Figure 14	Is the population of Planet Earth at a tipping point?	33
Figure 15	Planet Earth – the greenhouse effect	39
Figure 16	What are the impacts?	41
Figure 17	There are significant challenges in the Arctic as well!	42
Figure 18	Many of the world's major cities are on waterways which will be affected by rising sea levels	42
Figure 19	I'm telling you I've studied all the data and there is no waterfall around the bend …	43
Figure 20	The choices we make today will determine our and our children's future!	44
Figure 21	The challenge of setting CO_2 emission targets	45
Figure 22	Selected policies and measures to mitigate climate change	48
Figure 22	Selected policies and measures to mitigate climate changeconcluded	49
Figure 23	The ARROW approach to addressing Climate Change	51
Figure 24	You know these solar-powered clothes dryers work really well, and are very simple to use!	54
Figure 25	We really should have checked the oil before we put our foot on the accelerator!	57
Figure 26	US Energy 2008	58
Figure 27	I think we need to do some more work on the renewables!	59
Figure 28	I can't seem to get these numbers to add up!	60
Figure 29	Comparative emissions of sample journeys	66
Figure 30	The world's aquifers are running low!	69
Figure 31	Global fresh water usage estimates	70

Figure 32	I now pronounce the Dead Sea officially dead!	72
Figure 33	I can remember when the fish used to be this big!	72
Figure 34	Yes, we've got plenty of reserves!	74
Figure 35	Not tonight Josephine, it's too hot	79
Figure 36	Cumulative damage to the marine environment	80
Figure 37	The food/water bubble	81
Figure 38	The last tree?	89
Figure 39	Hey, did you know that the bees used to do all of this for nothing?	90
Figure 40	The treasures of our rainforests	92
Figure 41	The treasures of our rainforests are rapidly disappearing	93
Figure 42	So long, and thanks for all the fish…	93
Figure 43	We can rebuild the earth's forests!	95
Figure 44	Planet Earth – approximate wealth distribution	101
Figure 45	It's not working very well is it?	102
Figure 46	Let's not breathe today honey, let's drive. It's safer!	109
Figure 47	The interconnectivity of human health with the other key challenges	111
Figure 48	Give a person a fish and they eat for a day…, Teach a person to fish, and they eat for life …, Teach a person to read and they may end up running a global food supply company providing food to millions, funding reading programmes in other developing countries and sustainability programs for all of Planet Earth	119
Figure 49	I think there are a few pieces missing …	120
Figure 50	Mutual assured destruction syndrome (MAD)	127
Figure 51	The real costs of war!	129
Figure 52	The impact of climate change on security risks	133
Figure 53	You know, I really think they should have thought about the other challenges like climate change when allocating defence spending!	136
Figure 54	'They're beginning to get away, we'd better do something'	139
Figure 55	Estimated additional costs to address the basic social and earth restoration goals	146
Figure 56	Illustrative annual outlays to achieve the global goals	147
Figure 57	The interconnectivity of global warming and other Global Challenges	149
Figure 58	The interconnectivity of human health and other Global Challenges	151
Figure 59	The impacts of the Global Challenges on biodiversity	151
Figure 59	The impacts of the Global Challenges on biodiversity concluded	152
Figure 60	Impacts of climate change on water availability and potential conflict	153
Figure 61	I've got the solution! No, I've got the solution…	155
Figure 62	We've all got the solution!	156
Figure 63	The key agents in the Triangle of Change	167
Figure 64	No, it's their responsibility!	168

Figure 65	It's all of our responsibilties	168
Figure 66	We can all take part in building a sustainable world	183
Figure 67	The Business Planning Cycle	191
Figure 68	The Customer-Driven Business	194
Figure 69	Responding to external pressures	195
Figure 70	I think it's your turn to pull the ripcord! No, I think it's your turn!	198
Figure 71	The waves of Information Technology development	202
Figure 72	Trust between Chief Information Officer and Business Units	204
Figure 73	All I asked was did the customer like the product!	204
Figure 74	The key participants in the Triangle of Change	231
Figure 75	A problem of funding or a problem of management?	243
Figure 76	The Connected Reporting Framework	260
Figure 77	BT Group 2008 Values Report	261
Figure 78	Strategic Management for Managing Corporate Conflict Dynamics	270
Figure 79	Framework for types of business engagement	271
Figure 80	Corruption: Transparency International Six Step Process	274
Figure 81	The Ten Principles of the UN Global Compact	277
Figure 82	We've all got the solution!	282
Figure 83	The waves of change that have revolutionalised human civilisation	286
Figure 84	Delivering our Plan for the Planet	292
Figure 85	Plan for the Planet Prioritisation Matrix	293
Figure 86	Developing your own Plan for the Planet	294
Figure 87	Everyone can play their part in delivering our Plan for the Planet	295
Figure 88	One of the many reasons we should be working together on our Plan for the Planet	296
Figure 89	Planet Earth – It's the only home we've got!	297

List of Tables

Table 1	Direct greenhouse gas emissions of the typical UK individual	53
Table 2	Oil consumption for six key countries	62
Table 3	UK Energy consumption estimates	65
Table 4	Typical energy consumption of appliances within the UK household	65
Table 5	The business role in strengthening food value chains	85
Table 6	Largest military expenditures, 2006	129
Table 7	Significant ongoing armed conflicts, 2007	130
Table 8	The challenges of interconnectivity	150
Table 9	Solutions and cross benefits	154
Table 10	Typical relationships between values, vision and strategy	205
Table 11	Characteristics of static vs future-oriented organisations	208
Table 12	Partnering to strengthen public governance	249

About the Authors

Ian G. Chambers

Ian Chambers, M.B.A (AGSM), has worked in global corporations in the telecommunications industry over the last 20 years, in a range of executive roles including sales and marketing, business strategy and programme management. This has provided the frameworks for the development and deployment of global change management programmes – including corporate growth and sustainability strategies.

He has a Master's in business studies and an undergraduate degree in psychology, with his career also spanning business management in government, health and community action programmes.

He has in parallel been an active member of the Green Economics Institute Advisory Board and a member of the Academic Journals Advisory Board where his focus has been on establishing practical approaches to green business development and global sustainability.

John W. Humble

John Humble, an international management consultant, pioneered the practice of Management by Objectives, Service Management, the Social Responsibility Audit and the importance of Corporate Values. His six books have been translated into 17 languages.

He has a Master's degree from Cambridge University and is a Companion of the Chartered Management Institute, Fellow of the Institute of Business Consulting and Fellow of the Chartered Institute of Personnel Development.

His 6 film series on Management by Objectives (EMI, London) was internationally acclaimed and produced in 5 languages. John has also made 6 management practice films for BNA Films, USA.

In the past John has contributed to British national commitees, including terms of office with the Wool Textile Economic Development Committee, the Management, Education and Training and Development Committee of the National Economic Council and also the CEntrail Training Council.

John Humble's distinctions inclue The Ford Foundation Businessman Award, the Burnham Medal of the British Institute of Management, the Social Responsibility Award of Management Centre Europe and Fellowship of the International Academy of Management.

Acknowledgements

In building this plan, it is important to acknowledge the important contribution of many who have shown leadership and inspiration – and we have been privileged to draw on the wisdom and experience of these leaders who have been highlighting and developing practical responses to the global challenges. Lester R. Brown in his book *Plan B 3.0* (W.W.W. Norton Inc., 500, Fifth Avenue, New York, NY 10110, 2008) is insightful and definitive. Jeffrey D. Sach's books *The End of Poverty* (Penguin Books, 80, Strand, London, WC2R ORL) and *Common Wealth* (Allen Lane, Penguin Books, 80, Strand, London, WC2R ORL) provide a creative vision of the issues backed up by comprehensive research. Miriam Kennett and Volker Helleman through the *Green Economic Institute Journal* have progressed the debate on both the issues and solutions for moving forward as we embrace the new green revolution. Peter Dicken in his book *Global Shift: Reshaping the Global Economic Map in the 21st Century*, and Grazia Ietto-Gillies in her book *Transnational Corporations and International Production* have both expanded our horizons in understanding the global economic paradigms which will need to be addressed as we embark on this exciting revolution. We thank them sincerely.

There are many others who have given the world a 'wakeup call' through their pioneering studies and commitment. To name everyone is impossible but special thanks must go to HRH The Prince of Wales, Al Gore, George Monbiot, The World Bank, the UN and its subsidiary organisations, the World Economic Forum, the World Bank, Dan Smith, Transparency International, Optimum Population Trust and the World Business Council for Sustainable Development (WBCSD. Louise Kemp has been superb in converting concepts into the illustrations in the book. Paula Chaplin, our Research Assistant, has not only kept us on track with detailed information but has also made valuable broader contributions.

Ian Chambers would like to particularly thank the many friends and family who have been an inspiration and tremendous support in developing and delivering this Plan for our Planet. Graeme, Barbara, Emily and Melanie Chambers, Julie Skribins, Helen Boyle, Gundula Azeez, Howard Ford, Mike George, Deborah and Brian Gray, Neil Harris, Cathy Hewitt, Hazel, Kaye and Bob Hopping, Roger Jeynes, Nicola Kent, Barry and Judy Kinnaird, Jacqueline Rawls and Janine Whitaker. And finally, a special thank you to my parents Glad and Norman Chambers, who have always been a continual inspiration to build a better world.

John Humble wishes to thank Joan Watson for being an inspirational adviser, for her sustained commitment, and her rare gift of asking "out of the box" challenging questions. Last year I was seriously ill for many months. Joan was there for me, keeping my spirits high. Thank you Joan.

The 50% of royalties due to John Humble will be donated by him to the Humble Trust (Through Christ a friend for all seasons) CAF 16004482.

Finally, we thank those who have encouraged us to write this book and who share our passion for developing a successful Plan for the Planet. Miriam Kennet and the Green Economics Institute team, Sara Thomas, James Beaton, Ian and Anne Trott, Patricia Lustig, Martin Williams, Alan Cobb, Eric Rimmer, Bill Ryson, Jonathan Norman and all at Gower Publishing.

CHAPTER 1

Why We Need a Plan for the Planet

'The evidence is compelling: we need to redesign our social and economic policies before we wreck the planet.

At stake is humankind's one shot at a permanent bright future.'

Edward O. Wilson

'If a visitor were to arrive from outer space and looked at planet earth, their immediate request would be to ask to see the manager.'

William S. Burroughs

Let us just imagine for a moment that we did receive a visitor from outer space and after a brief look at the state of our planet, they did ask to see the manager! What would our response be? We could imagine taking them to one of our global organisations such as the United Nations. Here we could imagine that we would probably highlight the major advances our civilisation has made over the last 10,000 years – from hunter gatherers, to the wonders of modern science, medicine, communications and technology.

However, our visitor would probably not be totally satisfied with this overview and we could expect them to ask a few telling questions about what is not working so well on Planet Earth. Questions about climate change, energy resources, food and water, global health, unsustainable population growth and the well-being of other species on the planet.

At this point we would probably have to admit that we as a human race are facing some of the most challenging issues we have encountered as a civilisation – all converging at the same time and with many needing to be addressed in the next decade if we are to be successful in dealing with them.

We could imagine our visitor nodding understandingly and explaining that every planet faces challenges as it progresses. However the successful planets develop an overall plan on how to address the issues in a coordinated way and then work together to tackle them. At this point we know what the next question will be. Our interplanetary visitor would ask to see our 'overall plan' on how we are dealing with these challenges we are facing. How we are coordinating our efforts on a global scale? How we are working together to tackle these issues?

At this point we could imagine an embarrassed silence as we have to admit that we have no overall global plan to address all of these issues. In fact, we would have to admit that we are struggling to develop and agree a plan on even one issue such as climate change.

Now the realisation of the predicament we are in as a human race hits us. The human race and Planet Earth are facing some of the most challenging issues in the history of our civilisation, and yet we have no overall, coordinated global plan on how to deal with them.

This is even more surprising when we consider that we already know that an organisation or business which is facing major challenges will be much more successful if it has a plan of attack that everybody can get behind and implement.

Therefore we know how to put a global plan together and we know how to achieve global change. We even have examples where plans have been developed and successfully implemented to deal with global crises – such as the Marshall Plan following the Second World War – and more recently the coordinated action taken on the global financial crisis. And yet we have not translated these capabilities into an overall global plan for our planet to address the much larger crisis we face.

If this was a business, then our interplanetary visitor would probably suggest politely that we 'change the management' and appoint a new management team who can quickly develop and implement a plan of attack to address these global challenges.

However, at this point our second predicament hits us. The reality is that there are no other managers to appoint – we are 'the managers'. Everyday, in everything we do, we all manage a part of the 'business' called Planet Earth. This may be in our household, our business, in our community, in the organisations in which we work, or in government. And how we manage these parts of the business everyday impacts on Planet Earth – every one of us is everyday either making matters worse – or making things better. There are no bystanders. And there is no one else to address these issues. The reality is that it is up to each of us to make a difference in the areas we can impact.

However, we know from the basics of business and organisational management, that to be most effective, this can be most successfully achieved when done in a coordinated way. In fact, we don't need a degree in business and organisational management to know this. It is common sense. If we are going on a holiday, if we are moving house, we put a plan together. We don't just sit around and hope that it will happen because if we do this we know that it probably won't happen.

Yet, if we look at the challenges that we are facing on Planet Earth, we have to admit we have no overall plan that can enable us to tackle them in an effective and coordinated way. We are in danger of hoping things will sort themselves out, because it is unclear what we can do, and whether it will actually make a difference. In other areas, we have taken very positive actions, made developments and created best practices, but few would argue that we are lacking overall coordination. We have to admit we are simply not working effectively together to tackle and address the global challenges we are facing but at the same time, we are becoming increasingly aware of the extremely limited timeframes in which we have to act.

The time for just talking is therefore over; the time for coordinated action is now. We need to work together on a global as well as local basis to an overall plan. A plan that can be implemented at national, business, community and individual levels.

Is it possible to build and implement such a plan? We believe so, and have developed this Plan for the Planet as a starting point for exactly this. To demonstrate that using global business planning principles we can build a plan to achieve a sustainable world.

It is by no means a perfect plan. No plan ever is. However, it is intended as a starting point, to provide a framework which others can evolve, develop and adapt. We also hope it will provide the impetus to 'kick start' the actions and the changes that we need to make if we are to succeed in building a sustainable world. Not only for our generation but for all generations to come.

It is worthwhile at this point to take a few minutes to look at the changes that are taking place on Planet Earth, and why we urgently need a Plan for the Planet to ensure that we successfully implement it. These can be captured quickly by looking at the changes that will take place even over the next hour, the time it takes you to read a few chapters of this book. Within this next hour, the following changes will have taken place on Planet Earth:

- **Population:** Within this hour the population on Planet Earth will have increased by more than 9,000 people. This translates into an increase of an additional 220,000 people every day – and over 80 million more people every year![1] Over the last 70 years, less than one average lifetime, the human race has more than trebled from 2 billion to over 6.8 billion people.
- **Climate Change:** Within this hour over 1 million extra tons of CO_2 will have been released into the atmosphere – translating into over 10 billion tons per year.[2] The increasing evidence is that this is contibuting significantly to the damaging greenhouse effect and therefore the potential for further global warming. Global warming in turn has been linked to reduced crop yields, changes in weather patterns, and increasing droughts and storms. Events which we have already been observing over the last decade.
- **Energy:** Within this hour an additional 3.5 million barrels of oil will have been used. This means we continue to consume more than 30 billion barrels a year,[3] yet we are aware that at least half of the easily accessible oil reserves may have already been consumed. This is contributing to increasing uncertainty about long-term production capabilities. It has been estimated that Planet Earth without fossil fuels could only support 2 billion people, due to the importance they play in agriculture and food production.
- **Water and food:** Within this hour over 1 billion people will not have had enough food to eat or access to safe drinking water.[4] Women and children in Sub-Saharan Africa will be spending on average more than two hours a day collecting water,[5] with journeys of six to seven hours not unusual. At the same time, the European Union (EU) has estimated that over 40 per cent of its water resources[6] are being wasted.
- **Global resources:** Within this hour an area the size of 900 football fields will have been destroyed in the Amazon Rainforest – a major source of oxygen production for

---

1   As reported on the Wikipedia website: http://en.wikipedia.org/wiki/Population_growth.

2   As reported in Global Carbon Project (2008) Carbon Budget and Trends. Website: www.globalcarbonproject.org. 26 September, 2008.

3   As reported on the Wikipedia website: http://en.wikipedia.org/wiki/Peak_oil.

4   As reported in The Atlas of Water, Clarke R. and King J. Produced for Earthscan by Myriad Editions Ltd., 59 Lansdown Place, Brighton, BN3 1FL, UK. Website: http://www.myriadeditions.com.

5   As reported on the Council on Foreign Relations website: http://www.cfr.org/publication/11240/water_stress_in_subsaharan_africa.html.

6   As reported on the European Parliament website: http://www.europarl.europa.eu/sides/getDoc.do?language=EN&reference=A6-0362/2008.

Planet Earth.[7] In South East Asia and in the virgin Siberian forests[8] there is similar devastation. Forests that have stood for millions of years are being destroyed at an alarming rate. Financially, the EU has estimated that the total cost of environmental damage is more $8 million[9] every hour. However, we are all aware that the impact is much more than economic. It is devastating other species who also call Planet Earth home. More than 100 species are currently on the endangered species list.

- **Extreme poverty:** Within this hour more than 1 billion people will be struggling to survive on less than $1 a day and a further 2.5 billion people will be living on $2 a day.[10] Within this hour, over $300,000 of International Development Aid will be provided to try to tackle this inequity, yet less than half of this funding will reach those for whom it was intended.[11]
- **Global health:** Within this hour over 250 people will have been infected with HIV/Aids,[12] however, only 20 per cent of these people will have access to treatment. More than 200 young children will have died from diseases associated with poor hygiene and lack of sanitation,[13] and over 50 women in Sub-Saharan Africa and Asia will have died from preventable complications during pregnancy or childbirth.[14] Over 350 people will have died of tobacco-related illnesses.
- **Universal education:** Within this hour over 70 million children will still not have access even to basic primary education[15] and there will be over 800 million illiterate people, many of whom will struggle to find work.[16]
- **Conflict and peace:** Within this hour a further $130 million will have been spent on global military expenditure – contributing to a total cost of over $1200 billion per year.[17] In contrast, United Nations (UN) expenditure within this same hour on Peacekeeping will be less than $1 million, despite the fact that there are currently at least 18 significant unresolved conflicts taking place[18] on Planet Earth. Global expenditure on Peacekeeping is therefore less than $7 billion a year, compared to more than $1200 billion military expenditure.
- **Global finances:** Global GDP turnover will be over $70 billion within this hour, or over $650,000 billion ($65 trillion)[19] per year – arguably more than sufficient funds to address the global issues we have just reviewed.

---

7   As reported on the The Princes Rainforest Project. Rainforest is being destroyed at the rate of a football pitch every four seconds. www.rainforestsos.org.

8   As reported on the BBC News website: http://news.bbc.co.uk/1/hi/programmes/crossing_continents/7338623.stm.

9   As reported in the Brisbane Times. Website: http://www.brisbanetimes.com.au/news/world/the-cost-of-environmental-damage-5000000000000/2008/05/30/1211654255794.html, European Commission Report on Environmental Damage, 2008.

10 As reported on the Globalissues website: http://www.globalissues.org/article/26/poverty-facts-and-stats.

11 As reported on the BBC News website: http://news.bbc.co.uk/2/hi/in_depth/7622275.stm, Billions 'wasted' by aid system.

12 As reported on the NewScientist website: http://www.newscientist.com/article/dn9244-unaids-hiv-infection-rate-has-stabilised-at-last-.html, UNAIDS: HIV infection rate has stabilised at last, 30 May, 2006. , and UNFPA website: www.unfpa.org/swp/2004/english/ch1/index.htm - 26k P 7.

13 As reported in Time Magazine, 16 June, 2008. Website: http://www.unicef.org/health/index_problem.html.

14 As reported on the VOA News website: http://www.voanews.com/tibetan/archive/2007-12/2007-12-10-voa5.cfm, UN reports fewer children dying from preventable causes, Schlein L., Millennium Development Goals 2007 Report, p. 17.

15 As reported on the IHT website: http://www.iht.com/articles/2008/11/25/africa/school.php, Millions of children lack schooling, UN report finds, New York Times, Published: Saturday, October 25, 2008.

16 As reported on http://portal.unesco.org/education/en/ev.php-URL_ID=12874&URL_DO=DO_TOPIC&URL_SECTION=201.html.

17 As reported on http://en.wikipedia.org/wiki/List_of_countries_and_federations_by_military_expenditures.

18 As reported on the Ploughshares website: http://www.ploughshares.ca/ and news sources, Project Ploughshare.

19 As reported on the Wikipedia website: http://en.wikipedia.org/wiki/World_economy.

These examples give a sense of the scale of the global challenges we are facing on Planet Earth: Unsustainable Population Growth; Climate Change; Energy Supplies; Water and Food Supplies; Planet Sustainability and Biodiversity; Extreme Poverty; Global Health; Universal Education; Conflict and Peace and Financing a Sustainable World.

As we, however, review these challenges, the scale of investment, the coordinated global action required to address them and the urgency with which they must now be addressed appears daunting. The issue of Climate Change has moved from 'is it real' to 'is it too late' in less than five years. Many of the other global challenges such as developing alternative energy supplies, addressing increasing food and water shortages, and our devastation of the planet's biodiversity resources are equally urgent. The question also remains – is there the required political will, commitment and capability to tackle these global challenges?

The good news is that from the time we started walking upright the human race has demonstrated a great capacity for creative response to the challenges it has been confronted with on Planet Earth – to adapt. As the British historian Arnold Toynbee said, 'The quality of human nature on which we must pin our hopes is its proven adaptability.'

We already are seeing good examples of recent successes and a growing awareness that change is urgently required:

- **The ozone layer**: In the early 1970s, it was identified that the Earth's ozone layer, which protects all life from the sun's harmful radiation, was being damaged by man-made CFCs. Over a 20-year period a coordinated global plan to minimise the problem was developed and implemented. New products, free from CFCs, are now globally used and the damage to the ozone layer is repairing.
- **The eradication of polio**: In 1985 polio was killing or paralysing more than 100 million children a year in developing countries. Rotary International, driven by its mission to serve others, accepted the challenge and with the advice of Dr Sabin, developer of the oral polio vaccine, established a global strategy named PolioPlus. In partnership with health agencies, national health ministries and philanthropic organisations, and with a million Rotary members working as field workers, over 2 billion children were vaccinated within a 12-year period. Polio, widespread in 125 countries in 1985, exists now in only 20 countries. The mission continues. This is a clear demonstration that when concerned people commit themselves and act to a coordinated plan, they can make a real difference.
- **Increasing global awareness and coordination**: Over the last decade there has been an increasing recognition and understanding of the global challenges we are facing at every level and improved coordinated action to address them.

    The UN has given leadership through the Global Compact and the Millennium Development Goals which set specific targets to the year 2015, addressing a number of the global challenges. The Prince of Wales has established important programmes to improve issues such as deforestation and Climate Change, drawing on the leadership and funding of major businesses. The World Economic Forum, the Organization for Economic Cooperation and Development (OECD), The World Bank, the Green Economics Institute and many more are deeply involved. Indifference has given way to new initiatives and exciting 'on-the-ground' initiatives – as well as increasingly coordinated global action.

- **The business contribution**: Businesses are increasingly seeing these challenges no longer as threats, but as opportunities presented by the 'new green revolution'. There has been increasing investment in research and development in 'green' products and technologies, solar energy, improved control of emissions in coal power stations, new seeds and irrigation techniques and hybrid automobiles. The list grows daily. The influence of mobile technology and the Internet is already playing a significant role in many of the solutions to these global challenges and increasingly our ability to mobilise and coordinate on a global scale.

As we learn to adapt to the world which now confronts us, these are important starting points, and much progress is being made. However, there is also an increasing recognition that much more needs to be done, and quickly, if the challenges are to be effectively addressed within the extremely limited timeframes available.

Climate Change is taking place faster than initially anticipated, and needs to be properly managed within the next decade if we are to avoid the potential disasters of an overheated planet. Potential food, water and energy shortages similarly need to be addressed urgently, as does population, which continues to increase at an unsustainable rate. These factors in turn are contributing to the rate of forest, wildlife and biodiversity destruction which must be stopped before there are no more resources to destroy and there is no inheritance to pass on to future generations.

We need to begin acting a lot smarter and faster. The greatest challenge we therefore face, whether in government, in business or in the community, is 'how can we achieve this?' How can we achieve the 'step change acceleration' required to drive the coordinated changes across the planet? Can we work together on a global scale to implement this coordinated action, delivered at a local level, and in the limited timeframes required to avert the potential devastating impacts on the Planet Earth of the global challenges? Can we combine our efforts and actions fast enough to deal with these challenges before they combine together into what some are already terming the 'perfect storm'?

The Purpose of This Book

Fortunately, the answer is yes, and that it the purpose of this book. We outline an approach to achieving this step change acceleration through a coordinated planning approach which can be implemented at every level – government, business, community and as individuals. There is no doubt that the issues facing Planet Earth are challenging. Are they insurmountable? No. The fact is that many of these issues have been on the global agenda for decades and so have many of the solutions.

The biggest challenge has been in putting in place effective decision-making and management approaches to deal with the challenges and manage and coordinate their solutions on a global scale. Too many times we have dealt with each challenge on an issue by issue basis, with only limited success. Too often we have approached the challenges with a 'silo' mentality and not stood back to look at the bigger picture.

Lester Brown highlights the requirement: 'It seems that the world needs some sort of conceptual framework within which to look at the problems and then what to do about them. One would think that someone would be doing a plan – the World Bank or the

US Government or the UN. They're not even close. And you have to incorporate a wide range of issues because there are no partial solutions.'[20]

Again, the good news is, as well as having many of the solutions to these challenges at hand, we also have the required 'conceptual frameworks' available to address them. Business and many successful government and non-government organisations have been using global business management approaches to deliver global solutions using a coherent and integrated business-planning approach.

This approach is equally applicable to the challenges we are now facing, allowing us to look at the bigger picture, to understand how these challenges and their solutions interconnect, and to take a top down, globally coordinated, and locally delivered business-planning approach to the implementation of these solutions. That is the essential purpose of this book – to provide a business plan framework to the global challenges we face, and the coordination of the solutions which will enable us build a sustainable world.

Clearly a plan of this scale cannot be comprehensive. However it can provide an important starting point and conceptual framework for more coordinated and effective action.

The purpose of this book is also to provide a summary of the key challenges so they can be easily understood and to outline some of their solutions. This is to enable us to all understand not only the urgency for taking action, but also to demonstrate our belief that if we commit ourselves to using known solutions and best practices – leverage our ability to innovate and adapt – and work together to implement these capabilities in a coordinated and cooperative manner – we can build a sustainable future – not only for our generation, but for all future generations.

To build this plan, there are a number of key requirements:

- an understanding of the global challenges and solutions;
- an understanding of the interconnectivity of the challenges – and both the threats and the opportunities that these interconnectivities provide;
- an understanding of the key global management practices which can be used to develop and successfully implement a plan of this scope and scale;
- and finally the involvement, cooperation, commitment and sense of urgency and drive from all stakeholders – international and national governments, business and Non-Governmental Organisations (NGOs), communities and every individual in the community – to drive forward the necessary changes required to deliver our Plan for the Planet and build a sustainable world.

The book has therefore been structured to address these requirements – summarised in the following road map which outlines the issues and the sequence of steps needed to prepare and implement this coordinated global plan.

We cannot underestimate the challenge in implementing this road map on a global scale. However, we are greatly assisted by an understanding of the five key themes that run through this book:

- **Interconnectivity**: As we have already seen, many, if not all of the challenges are interconnected. The downside of this is that many of the challenges compound to

---

20   As reported in Scientific American, Vol 19, No 2, Summer, 2009.

A Road Map to a Sustainable World

Understanding the Ten Global Challenges

1. Population Growth
2. Climate Change
3. Energy Supplies
4. Water & Food Supplies
5. Planet Sustainability & Biodiversity
6. Extreme Poverty
7. Global Health
8. Universal Education
9. Conflict & Peace
10. Financing a Sustainable World.

Understanding the Interconnectivity of the Challenges

- The Challenges are interconnected and therefore cannot be solved separately.
- The timeframes to address the challenges are also limited – there is enormous urgency to act now.
- Too little, too late will bring the "perfect storm".

Recognising the Good News

- The human race doesn't know everything, but there are centres of excellence, best practices, new technologies and funds to make global improvements and actions quickly.

Developing a Plan for the Planet

We have the global management capabilities to bring these together into a coordinated, integrated plan.
- Building a Vision
- Setting Objectives
- Developing Strategies
- Establishing Action Plans
- Monitoring progress and continuous improvement.

Figure 2   Road map to a sustainable world

Building on more Good News

The human race has the depth of global and local management knowledge and skills to implement a Plan for the Planet

Delivering Our Plan for the Planet

Driving the change and making it happen requires that as with the successful implementation of any business plan, everyone needs to be involved. That means all of us have an important part to play.
- Politicians – international, national and local
- Business and NGO's
- Charitable, faith groups and community organisations
- Media
- Families and individuals.
- People – consumers, investors, electorate

Leveraging our Passion for the Planet

- As human beings, we must undestand we are visitors on this Planet, not owners.
- The current generation therefore has a profound responsibility to leave a sustainable planet, not only for this generation, but for all future generations to come.

"The hottest places in hell are reserved for those who in times of great moral crisis maintain their neutrality"
Dante

Figure 2   Road map to a sustainable world *concluded*

make matters worse. Increasing population puts further pressure on food and water supplies, as well as increasing the amount of greenhouse gases being released into the atmosphere, impacting Climate Change. Climate Change further influences water shortages, which in turn reduces agricultural productivity and can potentially increase regional conflicts as water becomes an increasingly scarce resource. You can begin to see how the vicious circle of interconnectivity can lead to the perfect storm!

However, the reverse is also true and there are very positive impacts of this interconnectivity. Successfully addressing one of the challenges can have a positive impact on other challenges. A very good example is better education for women. Not only does better education for women contribute to poverty reduction, it is also a major contributor to meeting the challenges of population growth, as better educated women have fewer children. It also contributes to improved family health. Interconnectivity demands therefore that the challenges are tackled in a coordinated manner through a coordinated plan to maximise the impact of the solutions and minimise any negative impacts of interconnectivity.

There is also the important interconnectivity of the three major agents of change in our society – government, business and people. We no longer have the luxury of saying it's somebody else's responsibility to tackle the challenges – it's all of our responsibilities. However, to ensure this happens effectively requires everyone working together to a shared global vision, with a set of common objectives and values. This allows each of these agents of change – whether they are national governments, charities, NGOs, businesses and international organisations, families and individuals – to coordinate their actions more effectively at global, national and local levels.

- **The stakes are high**: We are at a profound point in human history. Never before, one could argue, have the stakes been so high. The future of the planet and human civilisation as we know it is potentially at stake – and the solutions must be found by this current generation. The global challenges will affect everyone if they not properly managed and solutions found. The stakes are too high for self-interest to dominate.

    Human civilisation therefore has the opportunity to leverage one of the major capabilities that has enabled successful adaptation to the changing environment on this planet – cooperation. And now, due to our advances in technology, we are able to do this on a global scale.

    We have the opportunity to put aside differences, share experience in an open way and respect and listen to other people's view. In simple terms: these challenges combined are a 'common enemy' and only by partnership and cooperation can the battle be won and can we, as the generation entrusted with addressing these challenges, leave our grandchildren and their children a sustainable, peaceful and fulfilling world. The rapid advance in our understanding of challenges such as Climate Change, biodiversity and rapid resource depletion mean that this is no longer an idealistic hope but the harsh reality if we are going to make a difference.

- **A focus on business**: This book focuses on a business framework for this Plan for the Planet as business has a key role to play. Although many decisions and policies are required at a political level, the private business sector has a unique role to play as the powerhouse of economic growth, technical innovation, technological transfer and financial investment.

    Even more important is that business has experience of successfully driving the global management processes, managing global organisational structures and leveraging the large-scale transformational change management approaches required to achieve a successful Plan for the Planet.

    If we are candid, the most important failure in responding to the challenges to date has been a lack of a global approach to dealing with the challenges and the coordinated management of the solutions. The challenges are known, many of the solutions are understood, but effective and integrated global management and coordination has been lacking. Using global management best practices to address these challenges therefore presents an important opportunity to achieve the step change acceleration needed to be successful.

These best practice approaches can also be adapted by government, communities and individuals – as well as business providing a further opportunity for increased cooperation.

- **Urgency**: It is impossible to overstate the urgency with which the human race needs to respond to and manage these great challenges. Time is not on our side and what is done or not done in the next decade will shape the long-term future of our planet and all who live upon it. As Dr Martin Luther King Jr. said, 'Over the bleached bones and jumbled residues of numerous civilisations are written the pathetic words: too late.'

  Important in tackling these challenges with urgency, is also understanding that we do not have to wait for 'perfect answers'. Through cooperation, the human race has the opportunity to act quickly by implementing and continuing to develop and build on best practices. For example, Spain and Germany have very successful best practices in solar power generation and usage. These solutions are not only technological, but are also political and social with government incentives driving the required changes and community attitudes. These best practices can be quickly adapted by other countries. We do not need to reinvent the wheel, as the rapid transfer of knowledge and experience has never been simpler in our global village. The requirement for urgency is there, but our ability to respond urgently has never been more achievable.

- **'Together we can'**: Cooperation as we have seen, however, will be the key. And again, our ability to harness the capabilities and energy of everyone on our planet to achieve this has never been more realisable. This is not just in the areas of technological advance, government and business. Billions of people voluntarily give their time and energy to faith groups, charities and community service. For them, responding to the global challenge is a moral or spiritual imperative. Of course the driving force for radical change is to achieve a sustainable economic and social structure. There is also a more subtle motivation. An inescapable responsibility and opportunity for people of conscience to build a world which we are proud to leave to our children and future generations.

You are also invited to visit the Plan-for-the-Planet.com website where you can download the Plan for the Planet Templates to adapt to your own organisation, business community and household, keep up to date with developments and access the management effectiveness checklists. These are all located at: www.plan-for-the-planet.com.

Finally, you are invited to join us on the journey that this Plan for the Planet lays before us: a journey that will be fraught with obstacles, yet a journey that can be achieved through doing what the human race has done well throughout our history – working together to overcome the challenges that we face to build a sustainable world for our and future generations.

Figure 3   Save Planet Earth – for all of us! It's the only home we've got!

PART 1

Understanding Our Current Situation

'No man is an island entire of itself; every man is a piece of the continent, a part of the main. If a clod be washed away by the sea, Europe is the less ... Any man's death diminishes me. Because I am involved in mankind. And therefore never send to know for whom the bell tolls: it tolls for thee.'

John Donne, 1624

Let's start with an overview of our current situation, and in the next chapter examine in more depth each of the ten major global challenges and some of the solutions which are already being developed and implemented. First, lets look at the good news, and review our progress so far as a human race.

From Flint Chip to Silicon Chip

Looked at from any angle, the human race has come a long way as a civilisation in the last 10,000 years. From flint-wielding hunter-gatherers to farming communities, to the point where we now have a civilisation that boasts major breakthroughs in food production, health, longevity and technology. The last 10,000 years has seen a true demonstration of the remarkable innovative and creative powers of the human race.

Figure 4   **From flint chip to silicon chip!**

14   Developing a Plan for the Planet

In the last 200 years, this innovation and creativity has accelerated at an even greater pace. The advent of aviation, a man on the moon and space exploration, silicon chip-equipped computers and computer technology, the Internet and global communications – all have contributed to an age where the human race has moved from a local village to the beginnings of a true global village. From flint chip to the silicon chip – the major advances in medical, food production, space travel and health are achievements of which the human race can be truly proud.

Looking back over history, the human race has shown an amazing ability to adapt, to cooperate, to develop, to create and to deliver when confronted with challenges whether they be environmental, social, cultural or spiritual. That's the good news and that's where it is important to start this review. Whatever challenges we faced in the past, the human race has been able to survive and prosper.

However, There is a Problem …

Actually there are several problems. The wonderful advances in civilisation and technology have created challenges on Planet Earth that, as we have already seen, threaten to undo the work of centuries.

These challenges are all converging right now. Their interconnectivity is creating what some are calling the 'perfect storm'. However, as we have already seen, this also creates the 'perfect opportunity' for the human race to 'grasp the nettle' and cooperate on an unprecedented scale to build a better, more sustainable world. We need to work together to create the means for survival to avoid this potential 'perfect storm'.

In looking at this, it is again important to look at the issues of interconnectivity. The issue of global warming is closely linked to the increased extraction and use of carbon-based fuels such as oil, coal and wood. The scientific community has already developed a consensus on this issue, and it has received wide publicity. However, equally important, but under appreciated, is the relationship between increasing population and global warming.

Figure 5   **The perfect storm?**

The Perfect Storm?

As the human population increases there are simply more people using carbon-based fuels, resulting in more carbon emissions and therefore more global warming. When global population triples during a single human lifetime – from 2.5 billion people in 1950, to over 6.8 billion in 2010 – then carbon emissions triple, bringing about the current crisis. The problem is compounded as carbon footprint increases when people move out of poverty, thus having more disposable income which currently generates more greenhouse gases.

Increasing population, as we have already seen, is also clearly impacting other resources such as land and water usage. More food, water and land are required to feed the increasing population. This causes increasing deforestation and desertification and steadily reduces the natural 'sponge' that trees provide to absorb CO_2.

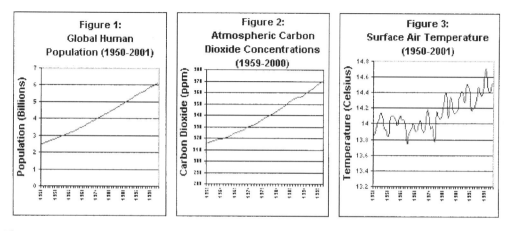

Figure 6   **A clear relationship between population growth, greenhouse gases and climate change**

Source: Diagram reproduced courtesy of the Population Resource Center. Website: http://www.prcdc.org.

Earth's vast oceans absorb huge amounts of CO_2. However, the levels of CO_2 in the atmosphere are now increasing at such a rate that as the oceans absorb more they are actually becoming more acidic – a phenomenon measured in the sudden changes to the population of certain aquatic species and the destruction of coral reefs.

The earth's oceans are also changed by the increasing levels of fresh water pouring in as glaciers and as the Arctic and Antarctic ice melts. This is reducing the salt content of the oceans, threatening not only marine life but also the flow of tides and currents. This could in turn affect weather patterns in regions such as Europe, through its influence on some of Planet Earth's weather control systems such as the Gulf Stream.

Water is also a resource under pressure due to population increase and the droughts created by Climate Change. Seventy percent of water usage is for agricultural production and that water scarcity clearly reduces food production, leading to shortages and potential conflicts that can exacerbate those shortages. We have also seen the movement away from carbon-based fuel to alternative fuels such as bio-fuels, produced from such crops as corn, and this is further increasing the pressures on dwindling land and food and water supplies.

Food shortages not only result from population increase but also from increased demand for food by populations newly lifted out of poverty. China is a notable example. An increasingly affluent population demands a greater supply of meat and other food resources. As this leads to more demand for animal feed, fewer crops are available to feed humans. Food prices are driven up for people still dependent on rice and corn. The recent food riots seen in developing nations in Africa, Asia and South America reflect the social consequences.

This also then raises social issues. Is it fair and reasonable to ask people in developing countries to stop the reduction in poverty levels and plans to increase levels of affluence to that of the European and North American populations, by reducing consumption and use of carbon-based fuels? This again highlights the importance of global cooperation to address the challenges. Action has to take place on a global, coordinated scale.

The rising scarcity of valuable resources also increases the likelihood of conflict. However, can we as a human race afford to engage in conflict, thus diverting scarce resource, manpower and technology, at a time when coordinated action is required to address the global challenges we are facing? Given the unprecedented destructive capabilities of modern weapons and the great concentration of populations in urban centres this could cause irreparable damage to our survival – at the very time when unprecedented cooperation is required to ensure this survival.

Finally, it is essential to consider the time factor. If we had 100 years to address these issues, then perhaps the traditional ways of negotiation and conflict resolution would be appropriate. However, scientists are already warning that global warming needs to be addressed comprehensively over the next 20 years if disaster is to be avoided. Shortages of water, food and energy are already becoming apparent on a global scale. Human population continues to grow, further accelerating the speed at which these changes are happening and the rate at which they must be addressed.

Crisis Can Also Create Opportunity

This chapter began with a summary of the massive advances that the human race has made as a civilisation: this is highlighted in that we, as a human race, have been able to adapt, innovate and cooperate. The news is therefore not all negative as the human race can use its innovative and cooperative capabilities to address these challenges. But we do not have time on our side. Action is often driven by a crisis or compelling event or events.

The human race now has that crisis, the compelling event or events driven by the prospect of the brewing 'perfect storm'. For the first time in human history there is an increasing recognition that action needs to be taken on a global cooperative scale. No longer can the majority of countries afford to agree on the terms of a global agreement such as the Kyoto Treaty, with progress delayed by the disagreement of a powerful few.

The outcome of the UN Bali and Copenhagen conferences on Climate Change in 2007 and 2009 provided hope that a more cooperative approach is possible. However, Bali and Copenhagen only tackled one of the global challenges being faced. The human race has recently seen a cooperative spirit and a willingness to put aside political and national agendas to address the global financial crisis which emerged in 2008. A similar approach has also been demonstrated in tackling the problem of ozone layer depletion and working together to address HIV/AIDS and polio. Cooperation is possible when the need is fully understood. At no other point in the history of human civilisation has a cooperative and coordinated approach to addressing the global challenges that we face as a civilisation and as a planet been more apparent.

A Word on Complacency, Fear and the Need for Cautious Optimism and Action

It is easy to state that we must cooperate in acknowledging the problems and work in unity to solve them, however –as we have seen already in the response to some of these global challenges over the last decade – the human response has at times been often confused and contradictory.

Unfortunately, complacency can often lead to a failure to face up to the hard facts. We often hear the following comments: 'It's all exaggerated by pressure groups. The scientists don't agree anyway', 'My way of life seems much the same now as it was. What's the fuss?', 'We have wonderful new technologies and skills which will overcome these problems', 'This is going to need sacrifice. Why should I be the first to volunteer?'

These attitudes, both at an individual and global level, can lead to a state of denial which, as we have seen on a global scale, paralyses constructive action. We have already seen this in areas such as Climate Change and global population stabilisation.

Fear can often leave people overwhelmed and paralysed – again resulting in a failure to take action. Attitudes such as, 'Clearly we face a disaster. It is too late to avoid or mitigate the consequences?' and 'We might as well live for the day' stop constructive action when critical intervention is required. This is why an overall plan which everyone understands and can play a part in is crucial, as it can show a way forward to address the challenges allowing success to build on success – overcoming the issues of complacency and fear, and replacing it with our requirement for moving forward rapidly – a cautiously optimistic approach.

Cautious optimism can lead to an understanding of the mainstream evidence for the large-scale problems which have to be faced, but also awareness that there are existing practical solutions and best practices which can be drawn on. Attitudes such as, 'There is no time to spare. We don't know all the answers but we know enough to make a big difference – so we can begin to act now', 'We need to educate and motivate the complacent and fearful people to join us', 'Everyone has a part to play' provide an important framework for action, in the face of sometimes overwhelming challenges.

A fundamental cause of inaction is also that many of the problems are complex and long term in nature with inevitable fluctuations in evidence even though the trend line is clear. Addressing the challenges of population and Climate Change are good examples. These issues have been known about and highlighted for over three decades. Yet we have seen procrastination and loss of urgency compared with responding to dramatic, acute crises such as the global financial crisis and short-term environmental issues such as the Tsunami.

We have, however, clear evidence of the danger of getting stuck in this way of thinking – failure to act can be blamed for the collapse of earlier human civilizations. It is important therefore to spend a few moments looking at what led to this collapse and the lessons we need to learn from these in order to tackle the global challenges we are currently facing.

18 Developing a Plan for the Planet

Fortunately, the reasons for the collapse of civilisations are captured well by Jared Diamond in his book *Why do Civilisations Collapse?* Sometimes they are destroyed by powerful neighbours or social and political changes. Many times, however, they have been driven by environmental changes, and a failure to address these, such as the degradation of the water supplies in the Fertile Crescent.

In the broadest sense, the global challenges we face now on Planet Earth are a combination of all of these factors – environmental, political, social and economic. So are there any lessons we can learn from these past civilisations which enable us to act and avoid a similar collapse?

Fortunately there are, and these can be best captured in the lessons we can learn from the experience of the small, isolated Pacific Island: Easter Island.

Easter Island Case Study – The Collapse of a Civilisation

On Easter Sunday 1722, Dutch Admiral Roggeveen landed on Easter Island in the Pacific, the world's remotest inhabited island, a mere 150 square miles in area. He found a few thousand people living in abject poverty, fighting one another over meagre food supplies and even resorting to cannibalism.

Yet, to his surprise, there was evidence of a once flourishing civilisation, with over 600 huge stone statues each 20 feet or so high. These statues could only have been produced by highly skilled craftsmen, who not only carved well, but knew how to transport and erect the statues, and with a social structure to support these craftsmen.

When we look back at the history of the Easter Island civilisation, we discover that there were a number of clans – each with a leader, who managed the community resources, religious and ceremonial activities, including the building of the impressive statues, which are a characteristic of the Easter Island we know today. The ocean was the main source of food, accessed by fishermen in their large canoes to what was seen to be an unending supply of nourishing fish and dolphins.

Figure 7 **Easter Island**

Understanding Our Current Situation   19

So what went wrong? Two key challenges that we are already familiar with led to the inevitable collapse of this complex and sophisticated community.

Firstly, as the population grew substantially, perhaps to as many as 15,000 people, the demand for crops and fish escalated.

Secondly, the rich forests and the island's other natural resources were steadily destroyed, in part due to the increasing population's need for canoes, buildings, fuel and other resources. A further major depletion came from the huge demands for timber to transport the giant statues from the quarries to their chosen sites. Large numbers of people dragged and guided the statues over tracks made of tree trunks.

We do not know whether the issue of deforestation was obvious to the people of Easter Island, however, we do know they continued to cut down trees until the last one had gone. Their cultural ambitions apparently blinded them to the threats to their quality of life. It could be argued that this is not dissimilar to what is happening on Planet Earth as a whole today.

Figure 8   I wonder whether he realises that he is cutting down the last tree?

The consequences were disastrous. Without timber, canoes could not be built and fishing became limited to onshore reed boats. Even the nets were no longer available because they had been made from the paper mulberry tree. Soil erosion led to reduced crops and the major source of food became chickens.

Many native birds, mammals, reptiles and plants were pushed to extinction through hunting or habitat destruction. Streams and springs dried up with the loss of the forests. The social structure broke down as rival clans fought for limited resources. The parallels, therefore, between the experiences of the Easter Islanders and our current situation on Planet Earth are clear. The challenges are similar: unsustainable population growth, limited energy, food and water supplies, major impacts on forests, land and other plant and animal biodiversity.

Similarly, we on Planet Earth have nowhere else to go. This is our only Planet, if the resources run out here we, like the Easter Islanders, have nowhere else to go.

Easter Island therefore is a timely reminder to us in microcosm of the significant consequences of over-exploiting limited essential resources and the failure to think ahead. So let's look at what lessons we can learn from the collapse of the Easter Island civilisation.

- **A lack of understanding of the impact of interconnectivity**: It is obvious, if we use our resources faster than they are replenished disaster is inevitable. When the Easter Islanders lost their forests they lost the capacity to fish, their land was degraded, fuel became a problem and the population was decimated by ill health. A classic case of interconnectivity. Yet, right now on Planet Earth we are destroying our forests, using up our limited water supplies and creating Climate Change. Our population continues to increase whilst our resources continue to decrease. The outcome is as obvious as in Easter Island – yet we do not have a coordinated global plan to address these issues. Whereas the Easter Islanders may not have had the understanding of this interconnectivity to make the necessary changes, our current generation does not have this excuse. We need to therefore ask ourselves what needs to be done to put a plan in place to make sure we do not follow the path of the Easter Islanders.
- **We need to anticipate the impact of population growth and resource usage**: We also need to recognise that the Easter Islanders were a clever people, who created a complex civilisation, and for many years a good quality of life. Why did they not anticipate the risk of deforestation and resource depletion? We however, have no such excuse. We have rich scientific knowledge; we know how to monitor weather patterns; we have great insights into the state of our planet, of health problems, of environmental issues and many of the solutions to these challenges. Ignorance is not a plausible reason for inaction.
- **Limited resources need to be effectively managed to avoid conflict**: As resources dwindled, bitter conflict arose between the different clans. The danger with this, as we have already seen, is that resources, manpower and technology – which could be focused on addressing the issues – are focused on addressing the conflict. Competition and argument between clans led to inaction. We have already seen this same situation delaying proactive action on Climate Change over the last decade. Further, we can already see how energy, water and other resource shortages are creating conflict in our world when the obvious solution is to work out equitable sharing plans. We need to learn the lessons from the experience on Easter Island, and recognise that a cooperative approach to understanding the challenges, putting a good game plan in place to address them, and implementing this plan is key to our success. We can only speculate as to the different world that would have been discovered on Easter Island had this approach been taken.
- **That means planning ahead on a global scale**: We can summarise that the Easter Islanders did not have the skills base and management abilities to anticipate their long-term requirements for sustainability and plan ahead. We have no excuse. We have a vast reservoir of knowledge and experience of problem analysis, setting objectives to address these problems, building management action plans and monitoring progress with creative information technologies. Why then are so many of the challenges we face not managed properly? Whatever reasons that may have existed until now, the requirement to rapidly and effectively deal with the global challenges means that good planning and management are critical for the survival of our civilisation.

Figure 9   A good case of bad planning?

- **And avoiding complacency**: Following a long-established way of life, taking for granted that resources would always be there, meant that the Easter Islanders apparently lost sight of the increasing urgency to deal with the growing risks. If we look at our approach to tackling many of the global challenges, we could argue that, similarly, we lack a sense of profound urgency – but without any excuse. We live in a globalised world where the challenges of conflict, global warming and population growth know no boundaries. We are increasingly dependent on one another. We will survive or decline together. Planet Earth is our only home. Like the Easter Islanders, we have nowhere else to go.

The lessons from Easter Island are clear. We need to look openly and honestly at the challenges that we are currently facing, cooperatively build and share solutions to these challenges – and most importantly – put a business-planning approach together that can enable us to tackle these issues in a coordinated and global manner. For the last of these we have drawn on the experience of global business practice in building this global planning approach.

However, identifying business and business practices as part of the solution will not rest easy with those who contend that the fundamental business models of industry and globalisation are to blame for the current crises. For this reason, it is important to spend a little time looking at why leveraging the skills, knowledge and capabilities of business are essential if we are to succeed in tackling the global challenges.

Why Take a Business-planning Approach?

Business and trade has been taking place for thousands of years as the human race has developed different skills, technologies, products and services that other people wanted and which therefore could be 'traded'. With this business practice, skills in innovation, wealth creation and management across borders developed.

This is the situation today, whether the transaction is a global deal involving millions of dollars, or a local activity providing services to people. Therefore, the basis of business brings important capabilities to the human race which can be quickly harnessed to address the challenges. Two of the most important of these are wealth creation and innovation.

However, there is also a third capability which is not immediately obvious, but which is increasingly important. Many of the management approaches developed to address business challenges and opportunities on a global scale provide important best practices that can help all of us – whether business, government, NGOs or people – in tackling the global challenges we face as a human race.

These 'Best Practice Management Approaches' that led to companies such as IBM, Coca Cola, Microsoft, Sony and Unilever successfully achieving their objectives and becoming global players in their business fields are available to us to use to address the challenges we face.

One obvious question that arises is, why use these types of best practice management approaches? The answer is simple, they are available now, they are proven, and they can be easily adapted to the current challenges that we are facing on Planet Earth. Innovators such as the Bill and Melinda Gates Foundation are already using these types of management approaches to tackle some of the global challenges. We will see how long it takes to develop solutions to areas such as Climate Change using conventional political approaches.

Time is no longer on our side. We therefore need to use our best management approaches, to help us fast track the development of planning of solutions and actions to address our global challenge. This approach has proven successful in providing the framework for this business plan for our sustainable world. The global management business-planning approach is therefore used to develop the framework for a global Plan for the Planet. Frameworks are also provided for the adaptation of these approaches to tackle the global challenges to each agent of change – whether that is government, business or people.

For the benefit of readers who may not be familiar with the business management approaches, brief summaries of these best practice global management approaches are provided. A series of checklists are also provided to allow readers to measure their performance against these best practices and develop strategies to maximise their performance.

Finally, the book finishes with an outline of how these best practices can be, and are already being, adapted to tackle some of the global challenges. Success in business requires involvement, cooperation and commitment from everyone. Therefore, the exciting prospect is that everyone, whether in business, in government or as a consumer, can get involved in tackling the challenges being faced on Planet Earth and can contribute to their solutions. The final chapter, therefore, highlights how the global management approaches can, and are being, used to address the global challenges – by businesses, by community groups, by NGOs, by individuals and by faith groups. We can all move from being part of the problem to being part of the solution!

A Word on Idealism

There are some people who argue that this cooperative approach of working together as a human race to address the global challenges is too idealistic. It would be good if it could happen but the human race has not always behaved in this way. With increasing population and dwindling resources is conflict over existing resources inevitable? Not necessarily, as the human race has survived and thrived by working together, whether hunting or farming, building and innovating. If we look at history, there have been much greater periods of cooperation than there have been of conflict.

Our civilisation is at a turning point with increasing demands and decreasing resources. Do we resort to increased conflict to meet the demands of a select few? Or as an alternative to increased global conflict, we can build greater global cooperation to increase the efficiency of existing resources, and develop new sustainable resources. This can be achieved – however, it can only happen if we are working together to implement a global plan.

It has, for instance, been estimated that if only 1 per cent of the solar energy reaching Planet Earth was harnessed then this could provide more than enough of the energy needs for the world's population. This is achievable in less than a lifetime. It has also been estimated that if the efficiency of solar power production increases at the same rate as computer technology has over the last 30 years, then solar power could provide for the energy requirements of the entire population of Planet Earth by 2050.[1] Focused, coordinated global action in this area could have enormous benefits.

There are many additional sources of renewable energy supplies being developed as well as solar. The solutions and best practices are there, the technologies are well advanced. The need to work together on a global scale to bring these solutions to fruition has never been so evident.

So let's summarise again how the rest of this book is structured to enable everyone to become part of the solution, not the problem:

Chapter 1: Understanding our Current Situation Defining the issues is a critical starting point in understanding each of the key challenges. There is so much information, sometimes apparently conflicting, that it is hard to identify the important issues.

Chapter 2: Understanding the Key Global Challenges For this reason Chapter 2 provides 'Executive Summaries' of the key global challenges being faced, some of the best practice solutions to each of these challenges, and how they can be applied by government, business and people to tackle these issues – and the issues and opportunities provided by their 'interconnectivity'.

Chapter 3: Developing a Plan for the Planet As we have seen, successfully meeting these challenges will require global coordination and effectiveness, and most of all, effective management. This can best be achieved by the development of an overall plan to which everybody can play a part in implementing. Chapter 3 sets out a framework for a Global Vision and Objectives and Strategies, a framework for a global Plan for the Planet.

Chapter 4: Managing a Plan for the Planet As some readers may be unfamiliar with these approaches, this chapter demonstrates how best management practice can make a vital contribution to developing a plan and putting it into practice.

Chapter 5: Delivering a Plan for the Planet The success of any plan can only be measured in its successful implementation. This chapter highlights how these business

---

1   As reported in *Earth: The Sequel*, Krupp, F. and Horn, M., W. W. Norton & Co. Inc, 500 Fifth Avenue, New York, NY 10110. Website: www.wwnorton.com.

management best practices can and are being used to address the global challenges by government, organisations, business and people – whether they be community or faith groups. The focus is dealing with examples and the dynamics of driving that change.

Planet Earth, like Easter Island, is an island sitting in the ocean of space. As we have already highlighted, there is nowhere else to go if this island is spoiled or made uninhabitable. At the moment we are heading in that direction.

However, the news is good. Never before have we had the technologies, the solutions, the global communication tools to mobilise everyone, and the management capabilities to manage the change that is required to save Planet Earth.

The purpose of this Plan for the Planet is to provide a business-planning framework for the development of a global vision, objectives, strategies and actions to enable us to achieve our vision – a sustainable Planet Earth and a better place to live for our children and all future generations.

PART II

Understanding the Key Global Challenges

'Man talks of a battle with nature, forgetting that if he won the battle he would be on the losing side.'

E.F. Schumacher

Understanding the Issues

It is important before we begin looking at plans and solutions that we have a good grasp on the major challenges faced by the human race:

- **Population growth**: Clearly, Planet Earth cannot support an unlimited number of people simply because its resources are limited. Yet the population growth over the last 60 years has been unprecedented. From 2.5 billion to over 6.8 billion and currently continuing to grow at an estimated 80 million people a year. Half of the existing population is under 24 years old – with significant fertility potential. There are increasing concerns that Planet Earth is already dangerously close to the human population limits and its resources are already being overtaxed.
- **Climate change**: The growth of human civilisation has taken place over the last 10,000 years in a period of relative climatic stability since the end of the last ice age. This stability has allowed the increase in food production, which has in turn allowed the rapid development of technologies. However, the stability of the climate is now changing due to human activity.
- **Sustainable energy supplies**: The sustainability of the energy sources that power human civilisation over the last 200 years are under threat. A major source of energy, particularly for food production and transportation, is oil. Yet the supplies of easily accessible oil are finite and cost-effective supplies may have already reached their peak production. The challenge of energy is therefore to develop alternative supplies which are not only in sustainable supply, but which won't further contribute to Climate Change within the limited timeframes available.

Figure 10   What on earth are we doing to our Planet?

- **Water and food supplies**: Population growth means more people placing demands on limited food and water supplies. There are already over 800 million people who do not have enough food to eat everyday, and over 1 billion with no access to clean drinking water. Coupled with this, the human race is obtaining a significant amount of its water supplies from non-renewable underground aquifers that have built up over thousands of years.[1] Increased food production requires increased water supplies yet these are declining. Clearly, this situation is not sustainable.
- **A sustainable planet**: Not only is the human race using up water and food supplies but other resources such as forests, lakes and fisheries are in dangerous decline. It takes a long time for these resources to replenish and some, such as wildlife and biodiversity resources, are irreplaceable. Some of the tropical rainforests in Asia and South America took over 300–400 million years to develop.[2]
- **Extreme poverty**: The increasing global population and the impact of energy shortages and Climate Change on food production means that there is an increased risk that even more people will not have enough to eat and will remain in extreme poverty. With over 800 million currently living on less than $1 a day,[3] and therefore not having enough to eat every day, this could have catastrophic consequences.
- **Global health**: Lack of adequate health facilities in developing counties has a number of consequences as well as the obvious ones of increased disease and death from common illnesses. In many of these countries, lack of healthcare results in people having more children to guarantee enough will survive to look after food production and to look after them if they are ill. It makes perfect sense in this

---

1   As reported in *The Atlas of Water*, Clarke, R. and King, J., p. 26. Produced for Earthscan by Myriad Editions Ltd., 59 Landsdown Place, Brighton, BN3 1FL, UK. Website: http://www.myriadeditions.com. As reported on the Council on Foreign Relations website:

2   As reported on the Godlikeproductions website: http://www.godlikeproductions.com/forum1/message376682/pg1, 300-million-year-old rain forest discovered.

3   As reported by The Royal Society and the U.S. National Academy of Sciences. Population Growth, resource competition and a sustainable World, United Nations, Global Monitoring Report, p. 2, 1992.

context but clearly exacerbates the population growth problem and helps maintain the poverty cycle.
- **Universal education**: A key issue that must be addressed when considering population, extreme poverty and health is access to education. Lack of basic education usually commits people to larger family size, poverty and reduced health. Therefore, access to basic education is the cornerstone of any actions to manage the Global Challenges.
- **Managing conflict and peace**: Peace, as already highlighted, is a key pillar in the development of a sustainable world. Conflict always dominates public concern and media interest and all too often valuable resources. For example, the 2007 G8 Summit, with its agenda tackling Climate Change, AIDS and pollution, was overshadowed by the escalation that was taking place between the US and Russia regarding the US plans to put 'deterrent missile bases' in Eastern Europe. Urgent global agendas such as Climate Change are often sidelined by conflict and war.
- **Financing a sustainable world**: In tackling any of the issues discussed so far there will be a financial cost – the price we will need to pay for building a sustainable planet. Identifying and sourcing funding for sustainability is therefore essential as is the professional management of cost effectiveness.

Executive Briefs

One of the challenges in understanding our global situation is getting good handle on each of the challenges as they are so diverse. For this reason, each of these Global Challenges is presented as an Executive Brief so the reader can quickly capture the issues and the opportunities. For readers familiar with an issue, a brief summary is presented at the beginning of each section. For readers who would like a greater understanding of particular challenges and potential solutions, the rest of the section provides a summary of the key aspects of each of these challenges.

The structure for each brief is:

- Summary
- The current situation
- Opportunities and best practices in tackling the challenge
- The role of Government, Business and People in tackling each challenge.

CHAPTER 2

Executive Brief No. 1: Population Growth

'If current predictions of population growth prove accurate and patterns of human activity on the planet remain unchanged, science and technology may not be able to prevent the irreversible degradation of the environment and continued poverty for much of the world.'

The Royal Society & USNAS (1992)

Figure 11   How many people can Planet Earth sustain?

SUMMARY

The issue of how many people Planet Earth can sustain has been hotly debated. The fact that Planet Earth is a finite size, with finite resources, has become increasingly apparent to all as the availability and degradation of these resources is taking place at an increasing pace. The human population may be at its own tipping point.

Unrestrained population growth on Planet Earth is unsustainable. In many places, including Africa and some parts of South East Asia, this population growth is as a result of lack of knowledge of family planning techniques, female education opportunities and availability of family planning resources. Population growth is uneven. Many European countries such as Russia and Germany are experiencing population declines which may lead to the rethinking of immigration policies.

Objectives based on the UN forecasts and strategies can be developed and rolled out globally by leveraging best practices from countries such as Iran, which have been able to reduce rapid

> population growth to sustainable levels. This has been achieved by the coordinated use of family planning and community education programmes, backed up by deployment of family planning resources to communities. A population increase from its present size of 6.8 billion to no more than 8 billion by 2050 is the objective.
>
> International governments and organisations have an important role to play in coordinating these efforts and focusing on the priority areas; national government in effective and efficient deployment; business in both provision of resources and education in workplace environments; and the community in adaptation of programmes to suit local and cultural requirements.

The Current Situation

- **Population has trebled in a lifetime:** Over the last 60 years the human population on Planet Earth has trebled from 2.5 billion people in 1950 to 6.8 billion people in 2009. The current estimates are that world population is projected to reach 7 billion by early 2012, and surpass 9 billion by 2050.[1]
- **Half of this population on Planet Earth is under 25 years old.**[2]
- **Growth is forecast to continue:** The current population is increasing at an estimated 80 million or 1.2 per cent per year. This translates roughly into 221,000 extra people every day, 9,200 extra people each hour and 2.6 extra people per second.

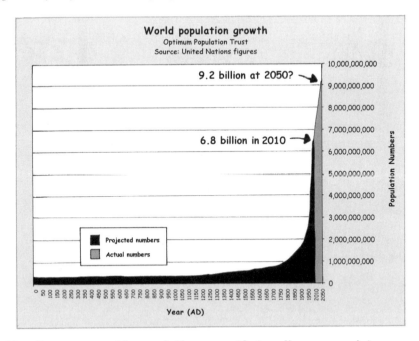

Figure 12 Forecast world population growth (medium scenario)
Source: Used with permission Optimum Population Trust.

---

[1] As reported in *The New York Times*, 11 March, 2009, UN: Young and Old Boom on the Road to 9 Billion (UN Population Division/DESA).

[2] As reported on the ABC website: http://www.abc.net.au/worldtoday/content/2003/s962527.htm. Half the world's population under 25: UN report, The World Today, Wednesday 8 October, 2003.

- **... even at the low fertility rates**: Even a reduction of one child per two families below the current forecast, a 'low-level' forecast, would result in a global population of approximately 8 billion by 2050. Therefore, continued population growth until 2050 is inevitable even if the fertility rates decrease. The UN has three estimates of future growth based on low, medium and high fertility rates.[3]
    - **Low Forecast – 8 billion**: The objectives set for the Plan for the Planet are based on the UN Low Forecast of 8 billion by 2050.
    - **Medium Forecast – 9.2 billion**: Based on the 'medium-level' fertility rate forecast, by 2050 the human population is forecast to increase by another 30 per cent to over 9.2 billion people.
    - **High Forecast – 10.8 billion**: Based on the 'high-level' fertility rate forecast, by 2050 the human population is forecast to reach 10.8 billion.

  It is worth noting that, since 2000, population growth has been closer to the 'high' level of growth than the 'medium'-level forecasts.
- **Most growth will occur in developing countries**: The population of most developed countries such as Germany, Italy and Japan is expected to reduce or remain almost unchanged. Russia is a dramatic example where a population decrease of some 22 per cent is forecast by 2050 due to low birth rates and increased death rates in a rapidly aging population.

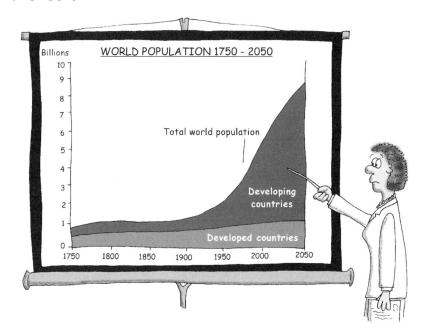

Figure 13   I think I know where we might need to place a little focus!

However, most underdeveloped countries are forecast to increase their populations by almost 50 per cent. By 2050, developing countries are forecast to add 2.3 billion people, compared to developed regions which are forecast to grow from 1.23 billion

---

3   As reported on the IIASA website: http://www.iiasa.ac.at/Research/POP/proj07/. 2007. Update of probabilistic world population projections and the UN Population Division Database: http://esa.un.org/unpp/.

to 1.28 billion. The growth pattern in developing countries is driven by migration from developing to developed countries, which is projected to average 2.4 million people annually from 2009 to 2050.
- **Lack of access to family planning resources**: Despite the increasing population challenge, resources to address the challenge are inadequate. Aid donors agreed to provide $6.1 billion a year for population and reproductive health programmes in 2005, but this was only one-third of the total needs.[4] As a result there are at least 200 million women still without access to effective and affordable family planning services. In many places funding has been declining in recent years. This is graphically highlighted in the Philippines, where cuts in US aid for family planning contributed to both unwanted population growth and the maintenance of poverty in poor regions.[5]
- **Current growth rates are unsustainable**: The UN states that if the current rates of growth continue until 2300, then there would be over 100 billion people on Planet Earth.[6]
- **An ideal population size?** The Human Footprint Team, after reviewing the earths capacity to support the human race, has calculated that the world as a whole, is overpopulated by two billion – the difference between its actual population and the number it can support sustainably, given current lifestyles and technologies[7] This raises major political, economic and social questions.

IMPACTS

- **Decreasing resources, increasing population**: These rapid increases in population are already placing Planet Earth under increasing stress from human activities, with its climate changing and key ecosystems failing. Countries such as Egypt, Ethiopia and Sudan are competing for the resources of the Nile whilst their populations are forecast, to double in some cases, by 2050.
- **Fatal consequences for Planet Earth's ecosystems**: The uncomfortable truth is that the impact on the earth's biosphere of 9 billion people living at a desired higher standard of living in 2050 would be fatal for the planet in terms of greenhouse emissions alone.
- **Population increases linked to greenhouse gas emissions**: The rise in greenhouse gas concentrations in the earth's atmosphere, and associated temperature rises, follows the sudden and steep rise in population numbers from the start of industrialisation less than 300 years ago.
- **Increased migration**: Another key factor influencing country population size is migration. There were over 190 million international migrants in the world in 2000

---

4   As reported in Brown, L. *Plan B 3.0*, p. 120. Earth Policy Institute, W. W. Norton & Company. 500 Fifth Avenue, N.Y. 10110.

5   As reported on the Populationaction website: http://www.populationaction.org/Press_Room/Viewpoints_and_Statements/2008/04_24_Philippines.shtml.

6   As reported on the Populationpress website: http://www.populationpress.org/publication/2004-1-un.html, World Population in 2300 to be around 9 Billion.

7   The calculations have been made possible by advances in the methodology of ecological footprinting, which measures the area of biologically productive land and water required to produce the resources and absorb the waste of a given population or activity and expresses this in global hectares – hectares with world-average biological productivity. The index uses data contained in the latest Ecological Footprint Atlas, produced last year by the Global Footprint Network and based on figures for 2006. As reported by Population Trust. Website: http://www.optimumpopulation.org/opt.release18Feb08.htm. News release, February 18, 2008.

– one in every 35 people, and up from 79 million in 1960.[8] Migration is increasing from poor countries to rich countries. 13 per cent of the US population was born in another country. France has 6.5 million immigrants, equivalent to 11 per cent of the total population. Remittances – sending money back to their home countries by migrants – were almost $250 billion in 2004, mostly to developing countries. This is the total size of foreign aid. Migration from developing countries to developed countries also increases each individual's carbon footprint thus causing further impact on the environment.

- **Increased urbanisation**: For the first time in the history of the human race, the majority of the world's people are living in cities.[9] Developed countries saw permanent migration rise in 2005 at an annual rate of 10 per cent per annum. This issue is compounding as the number of people living in cities is growing twice as fast as overall population growth. In 2004, there were 20 mega cities with more than 10 million population. In Latin America, 77 per cent of people live in urban areas. There has been an alarming growth in slums with poverty and serious water and air pollution problems.

- **A strategic issue**: Continued population growth, long ignored as a strategic problem, is now being increasingly acknowledged to be unsustainable by world leaders.

Figure 14   Is the population of Planet Earth at a tipping point?

- **Aging population**: Only in the last decade has the issue of aging populations been fully acknowledged, yet it has potentially devastating economic and social implications. UN statistics show that 11 per cent of today's population of about 6.8

---

8   As reported by the UN Population Fund. Website: http://unfpa.org/pds/poverty.htm. and UNFPA State of World Population, 2004, p. 4.

9   As reported by the Economic and Social Research Council. Website: http://www.esrcsocietytoday.ac.uk/esrcinfocentre/about/ci/cp/our_society_today/news_articles_2007/half_mankind_in_cities.aspx?componentid=20871&sourcepageid=17746. More than Half of Population Living in Cities and UNFPA State of World Population 2004, p. 4.

billion are over the age of 60. By 2050, when the population will have grown to about 9 billion people, the proportion of over 60s will rise to 22 per cent in developed countries and 33 per cent in developing countries. With better healthcare, and improved diet, the average life expectancy has extended to 78 years in developed counties and 67 in the developing world.

By 2050, the UN predict the population will level out with women having fewer children. These changes mean there will be a heavy burden on young working people as they care for the elderly. The escalation in the cost of publically funded pensions will put a strain on a countries economic prosperity. Health facilities will be overstretched. Levels of savings will be reduced. Each country needs to prepare a strategic plan to meet these challenges.

Opportunities and Best Practice

OPPORTUNITIES

- **Universal access to reproductive health services**: Provision of services to the millions of couples who lack access to these services can significantly reduce the number of unwanted pregnancies. As the current number of unwanted pregnancies is estimated to be up to 80 million, which is roughly the equivalent of the estimated annual global population growth rate, effectively addressing the issue of unwanted pregnancies would potentially stabilise global population growth.
- **A focus on high population growth developing countries**: Developing countries have the highest forecast growth rates. Many of these countries have had the benefits of improved health programmes without the access to improved fertility control. UN growth statistics show the following top vulnerable ten countries: India, China, USA, Indonesia, Pakistan, Nigeria, Brazil, Bangladesh, Congo and the Philippines. Improved coordination and deployment of the provision of family planning education and resources to these countries could clearly have a significant impact.
- **Female education**: Increased education and empowerment of women has a direct correlation with reduced fertility rates and therefore similarly provides an important opportunity to address the challenge of global population growth.

BEST PRACTICE

- **Iran**: In the early 1990s, Iran developed an effective approach to stabilising population growth in a country that was previously experiencing high fertility rates. Population growth was halved between 1987 and 1994, to a level of 1.3 per cent, only slightly higher than the United States, through a national programme of family planning policies and education programmes, in a largely Muslim population (see box insert).[10]

---

10 As reported on the Earth Policy Institute website: http://earth-policy.org/Updates/Update4ss.htm. Iran's Birth Rate Plummeting at Record Pace: Success Provides a Model for Other Developing Countries. Larsen, J., December 28, 2001.

IRAN'S FAMILY PLANNING PROGRAMME

'Iran's success story in population and reproductive health is a vivid example of how political commitment, in its broadest sense, can bring about a sea change in development in a record time.'
Mohamed Abdel-Ahad, UNFPA Country Representative

The Challenge. Following the Islamic Revolution in 1979 and the war with Iraq in 1989, a large population was seen as a strategic advantage. Family planning clinics were dismantled and Ayatollah Khomeini encouraged high birth rates to increase the ranks of 'soldiers for Islam'. The population increased from 34 million in 1979 to 49 million in 1986.

However, it became clear that this growth was a great danger to Iran's future prosperity. There simply was not enough food, employment and infrastructure to cope. A major shift in policy took place.

Leadership commitment: Political and religious leaders committed themselves to a fundamental strategy of family planning, underpinned by legislation and supported by a wide range of new initiatives.

The Family Planning Bill: This Bill, introduced in 1993, removed many of the incentives for large families – some social benefits were only provided for the first three children. The Bill also planned for improved education and employment for women and social security improvements so parents would not be motivated to have many children as security for their old age.

Strategic action: A wide range of complementary actions were taken within this new strategy including:

- A two-day course in family planning before obtaining a marriage licence.
- Some 15,000 health clinics providing free of charge advice and supplies such as condoms, the pill, sterilisation and vasectomy. This scale of clinics was essential with a widespread rural population.
- A sustained and imaginative media campaign through TV, radio and the press.
- Improved education for women and girls through the Literacy Movement Organisation. This includes literacy courses, a range of vocational training, education on family planning and entrepreneurial skills. Two thousand community learning centres were established. Female literacy increased from 52.5 per cent in 1976 to 62 per cent in 2002 – representing a significant cultural shift towards greater gender equality.

Changes in population growth: As a result, there has been a dramatic decline in fertility from more than six children per woman in the early 1980s to just over two births per woman in 2005.

Wider issues: Iran is country faced with significant water shortages. The economic damage which would have been caused if the population growth had not been reduced would have been catastrophic. The population is now some 71 million but the forecast before the change in policy was as high as 100 million. However, this is still 'work in progress'. A major challenge now is to find employment for the large number of young Iranians who are approaching their reproductive years.

> Key lessons for the Global Challenge of growing populations from Iran:
> - A committed political and community leadership to addressing the challenge.
> - A legislative framework providing policy guidance and incentives.
> - A integrated comprehensive series of actions. Quick fixes don't work.
> - The education and empowerment of women.
> - A compassionate, learning-based approach.

The Role of Government, Business and People

Many of the countries with high population growth are also facing limitations in key resources such as family planning, water, food and land and there are not the funds to deal with these problems. This is where the three agents of change – Government, Business and People – can assist in coordinating improvements.

GOVERNMENT

- **Think global**: Population stabilisation needs to be coordinated at an international level to ensure the required resources are available to the nations requiring support, and at a national level to ensure that these resources are properly deployed. There is an important opportunity for the UN to increase its proactive role in monitoring and forecasting population growth, as well as providing consultancy in best practice. This can be provided through its agencies and funding sources for national action programmes, thus ensuring universal access to reproductive health services.
- **Act local**: Developed nations have the opportunity to ensure that the shortfalls in funds to developing countries are overcome to ensure provision of universal access to reproductive healthcare services. National governments in developing countries can ensure that all the key departments are involved and coordinated to effectively implement population stabilisation programmes. These include government departments covering education, culture, health and communications. It is also important to liaise and ensure effective coordination with both NGOs and businesses.

BUSINESS

Private business and NGOs can significantly contribute to a country's national population stabilisation in a number of ways:

- **Communication**: Businesses and NGOs can support the national communication strategy by explaining population stabilisation plans and how they can be implemented.
- **Education**: Businesses and NGOs play a vital role in education programmes for workers and the general population. Provision of family planning and health advice can be an integral part of these programmes in both the workplace and the community.

- **Healthcare**: Provision of healthcare and health education, particularly in developing nations, is an important contribution. The clinics and hospitals provide a further source of advice to workers and the local community on family planning and can act as contraceptive distribution points. There can be a double benefit since, when contraceptives such as condoms are used, they also reduce the risk of HIV/AIDS which is a serious cause of low productivity in the workforce.

PEOPLE

- **Local community support programmes**: The high mortality rate in children under the age of five means that many families in Sub-Saharan Africa are likely to continue to produce more children to guarantee support for food production, for illness and old age. The provision of community support programmes reduces this dependency, combined with programmes aimed at reducing under-five mortality rates to levels set in the Millennium Development Goals.[11] Local community support and sensible adaptation of best practice to suit local conditions and culture is an important way of leveraging international resources and support.
- **Healthcare and social security**: Provision of adequate social security systems and healthcare systems such as local free health clinics become key components of a successful population stabilisation programme.
- **Understanding**: Promotion of understanding of the challenge of unsustainable population growth is a key opportunity for all community sectors. This issue is often unreported and not linked to the other challenges, such as resource sustainability and Climate Change to take just two. Increased understanding and promotion of the issue of unsustainable population growth, along with provision of family planning education and resources to address it – are the keys to successfully addressing this Global Challenge.

> *Better education and empowering of women is one of the most effective ways of reducing fertility rates and therefore population growth.*

---

11   As reported on the Wikipedia website: http://en.wikipedia.org/wiki/Millennium_Development_Goals. Millennium Development Goals (number 4): Child Mortality.

CHAPTER 3

Executive Brief No. 2: Climate Change

'Man is trampled by the same forces he has created.'

Juana Frances

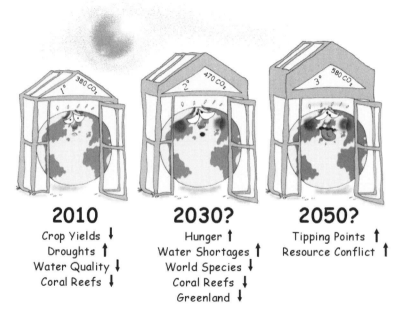

Figure 15 Planet Earth – the greenhouse effect

SUMMARY

The issue of Climate Change has moved quickly over the last five years from 'is it real?', to 'is it too late?' Leading scientists have identified the rapid rise in greenhouse gas concentrations – from 270 parts per million (ppm) of CO_2 in the atmosphere at the beginning of the Industrial Revolution, to over 380 ppm now – an increase of over 30 per cent as caused by human activity. These increased greenhouse gases in the atmosphere trap heat, and have raised the temperature on Plant Earth by almost one degree in this same period. Already the effects of even this one degree increase are being felt in changing weather patterns, melting polar icecaps and glaciers. This can potentially raise sea levels to catastrophic proportions.

> The actions required are twofold. Firstly, rapidly reduce and eventually eliminate greenhouse gases being put into the atmosphere. The EU and the US have both set targets of 20 per cent reduction by 2020, and 80 per cent by 2050. It is estimated that this would achieve a capping of the CO_2 in the atmosphere of 580 ppm CO_2 and three degrees of global warming. Two degrees, however, has been identified as the maximum that can be allowed before catastrophic and irreversible Climate Change begins to occur. The challenge is to get these reduction targets consistent across the globe – for developed and developing countries – and to then ensure that they are achieved. Government, Business and People all have an important role to play in achieving these objectives using the ARROW approach – Avoid, Replace, Reduce, Offset, Watch and Monitor.
>
> The second requirement is adaptation. Already, the impact of rising sea levels and weather patterns are increasing. As temperatures continue to rise, the need to implement plans for adaptation to these changes is critical.

The Current Situation

- **Temperatures are rising**: The average global temperature on Planet Earth has increased by 0.8°C from pre-industrial levels. 40 per cent of this temperature rise took over 150 years; while the final 60 per cent required only the last 60 years.[1] The temperature rises after 1975 is unprecedented, with the 11 highest recorded temperatures happening in the last 13 years.[2]
- **Temperature rises are linked to greenhouse gas emissions**: Temperature increases have been directly linked to a similar increase in CO_2 levels.
  - The atmospheric concentration of carbon is now over 380 parts per million – against 270 ppm before the industrial revolution.[3]
  - On current trends the figure will pass 450ppm within a decade and will be more than 580ppm by 2050 due to the increases in the emission of greenhouse gases by human beings.
  - Amplification of the earth's natural greenhouse effect by the build-up of greenhouse gases introduced by human activity has the potential to produce dramatic changes in climate.
- **Human activity is causing Climate Change**: In the twentieth century, the near quadrupling of human population and more than tripling of per capita CO_2 emissions of greenhouse gases has created a situation where it is now agreed that human activity is causing Climate Change.[4] The impacts in the following UN diagram show the range of consequences of Climate Change.
- **Increasing impacts**: As greenhouse gas levels and temperatures continue to rise, the damaging impacts on Planet Earth will increase.

---

1   As reported in Oil, Population and Global Warming, García, M. (Jr.). Website: http://www.swans.com/library/art10/mgarci10.html.
2   As reported on Science Daily website: Top 11 Warmest Years on Record Have All Been in Last 13 Years.
3   As reported in *The Atlas of Climate Change*, Dow, K. and Downing, T. E., Earthscan, Dunstan House, 14a St. Cross Street, London, EC1N 8XA.
4   As reported on the BBC News website: http://news.bbc.co.uk/1/hi/sci/tech/4969772.stm. 'Clear' human impact on climate.

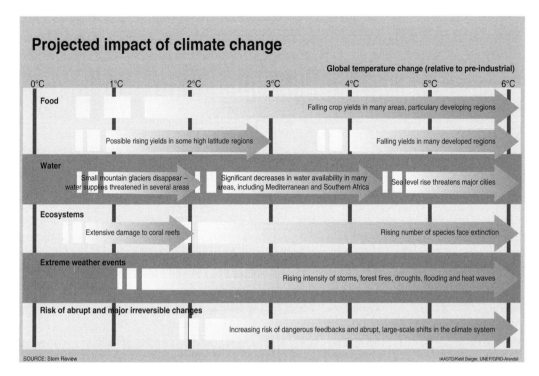

Figure 16   What are the impacts?
Source: Used with permission IAASTD/Ketill Berger, UNEP/GRID-Arundel.

- **Climate Change and the impact on Planet Earth's oceans**: The National Centre for Atmospheric Research estimates that 118 billion metric tons of carbon dioxide were absorbed by the earth's oceans between 1800 and 1994 – a major contribution to reducing the impact of greenhouse gases. However, this is causing the oceans to become more acidic as the dissolved carbon dioxide becomes carbonic acid – a corrosive agent which eats away the shell of important species in the global food chain. Coral reefs under attack may face extinction with inevitable damaging side effects on ecological balance. One hundred and fifty five marine scientists from 26 countries recently signed the Monaco Declaration, identifying the twin threats of global warming and ocean acidification as 'the challenge of the century'.
- **The Antarctic**: The British Antarctic Survey reported in 2009 that the peninsula has warmed by more than 3°C in the past 50 years. A 1 per cent loss of Antarctic land ice would probably raise sea levels by 65cm according to the Norwegian Polar Institute.
- **The tipping points**: There is also concern that the changes will not be simply incremental because if certain 'tipping points' are reached, then major catastrophic change such as sea level rises and collapse of the Gulf Stream which warms Europe could happen very quickly.[5] There is increasing concern that some of these tipping points may have already been reached.

---

5   As reported on the Wikipedia website: http://en.wikipedia.org/wiki/Tipping_point_(climatology).

Figure 17 There are significant challenges in the Arctic as well!

- **Managing the impacts**: Scientists around the world have also concluded that flooded cities, diminished food production and increasing storm damage all seem likely as a result of global warming and could affect the lives of billions of people. However, to date we have not taken significant action to address these potential impacts.[6]

Figure 18 Many of the world's major cities are on waterways which will be affected by rising sea levels

---

6   As reported on the UN News Centre website: http://www.un.org/apps/news/infocus/sgspeeches/search_full.asp?statID=121. Address to high-level event on Climate Change, Secretary-General Ban Ki-moon, 24 September, 2007.

Climate Change 43

- **Impact of industrialised nations**: Most of the man-made global warming pollution currently in the atmosphere has come from industrialised nations.
- **Impact of developing nations**: Emissions from major developing countries such as China, India and Brazil are increasing rapidly, with China currently overtaking the USA as the largest emitter of CO_2. If developing countries, with high population growth, follow the wasteful and inefficient western lifestyle model, the problems will clearly become worse.
- **Getting agreement to act**: Although the UN reports conclude that the majority of scientists agree that global warming is occurring and that it is caused by human activity, there is still significant confusion about Climate Change and the remedial actions required.

Figure 19 I'm telling you I've studied all the data and there is no waterfall around the bend ...

This has lead to significant 'analysis paralysis' when what is required is decisive and globally coordinated action and discussion as to who should start first. It is clearly in everybody's interests for a globally coordinated and rapidly deployed approach to greenhouse gas reduction to take place across both developed and developing nations.
- **Adaptation**: The impacts of Climate Change are already being felt – more destructive storms, rising sea levels and changing weather patterns. Despite this, emissions of fossil fuels continue to increase from an average of 1.3 per cent per year in the 1990s to more than 3 per cent per year from 2000 to 2008. Therefore, efforts to reduce greenhouse gas emissions need to be accompanied by proactive adaptation plans in areas such as energy, water, buildings and transport.

Opportunities and Best Practice

KEY CONSIDERATIONS

- **Alternative technologies are available**: Alternative non-greenhouse gas-emitting energy technologies are increasingly available. For example, solar, wind, geo thermal power and bio-fuels.

Figure 20 The choices we make today will determine our and our children's future!

- **Best practice precedents exist**: The reduction in the ozone layer is an example of success when scientists and governments work constructively together on a global basis to tackle a global environmental issue.
- **Progress can be made**: The identification of the challenge of acid rain led to coordinated political action for the introduction of emission controls on industry and a switch to natural gas (from coal) for power generation. The result has been significant declines in emissions of sulphur dioxide and nitrogen oxide emissions which are the gases largely to blame for creating acid rain. As a result of the halving of levels of acidic sulphur in British waters in the last 15 years, fish have begun to return to rivers and streams that were once acidic waterways.
- **Ambitious timeframes can be met**: The ability to develop and roll out new technologies quickly is illustrated by the fact that only 20 years after their introduction, over half the world's population, more than 3.5 billion people, use mobile phones. There are already over one billion computers and Internet users. Rapid deployment of technology is possible where focused, globally coordinated efforts are deployed.
- **Governments are starting to respond**: The EU's long-term strategy is that 20 per cent of energy must come from renewables and 10 per cent of transport fuels will be bio-fuels. EU emissions must be reduced to 20 per cent below 1990 levels by 2020. Many states in the USA are developing their own strategies. President Obama indicates the US could soon be leading the battle against Climate Change. He supports a Cap and Trade system to limit carbon dioxide emissions and has called for an overhaul of US industry which would require a $150 billion investment in renewable energy (see further review of Cap and Trade in The Role of Government section).
- **Preferred maximum increase in temperature**: It has been estimated that average global temperature rises could be between 2°C and 6°C – depending on how quickly the activities impacting the rises are curbed. The estimated maximum

temperature rise that Planet Earth can sustain without moving into this catastrophic change has been identified at 2°C.[7] It is important to note that this is not a 'safe' level of warming; it is merely 'less dangerous than what lies beyond'.[8] The need for urgent action on a global scale to minimise the temperature rises is therefore critical.

- **What greenhouse gas emission reductions are required to stay below the two degree threshold?** A number of estimates have been made about reduction levels and these lie in the range between 60 per cent and 90 per cent by 2050. The EU has set targets at 80 per cent reduction and 550 CO_2 levels. Environmentalist George Monbiot, in his book *Heat*, recommends a much more aggressive approach of 90 per cent by 2030 to stabilise at what could be viewed as a safer CO_2 level of 450 ppm to achieve a maximum of 2°C increase. It is essential that effective targets are agreed and rapid and coordinated global action is taken to achieve these reduction levels.
- **How fast should change be driven?** Achieving an effective balance between the risk of not changing fast enough, thereby increasing the chance of not achieving the 2°C degree maximum increase, and too fast, thereby not allowing the effective management of the changes required is crucial. An effective balance must therefore be achieved in driving to achieve these objectives.

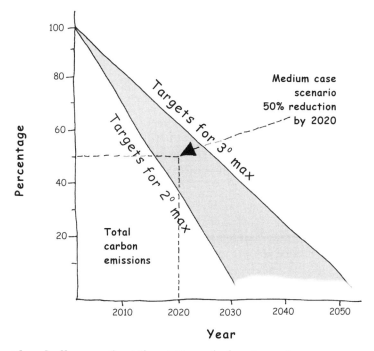

Figure 21 The challenge of setting CO_2 emission targets

OPPORTUNITIES

There are a wide range of activities and solutions which can be used to manage Climate Change. Below are some examples on which to focus for maximum impact to be made:

---

7 As reported in *Heat: How to Stop the Planet Burning*, Monbiot, G., p. 15, Allen Lane Publishers, 28 September, 2006.
8 As reported in *Heat: How to Stop the Planet Burning*, Monbiot, G., p. 15, Allen Lane Publishers, 28 September, 2006

- **Focus on high-impact countries**: The top 20 high greenhouse-emitting countries contribute over 70 per cent of current greenhouse gases and therefore are the top priority for action.[9] There has been significant debate about whether it is the responsibility of the developed nations, which have emitted most to the greenhouse gases to date, to take action, as they commenced their industrialisation much earlier; or the emerging developing nations to curb their rapidly increasing emissions. The answer is to focus on the countries which are currently the highest emitters of greenhouse gases – both developed and developing nations
- **Focus on forests**: Equally important is that 20 per cent of greenhouse gas emissions are coming from the burning of forests (that is, deforestation).[10] Therefore, added to this list must be the urgent need to address this problem in the key regions where large-scale deforestation is taking place: the Amazon, South East Asia and Russia.
- **Communication**: People change behaviour when they understand why it is important. Effective communication is required to enable people to understand why change is important, and that they can make a difference.
- **Make it easy**: Making it easy for people to make these changes is a key role that can be played through government education and legislation, as well as through innovative business products and services. This can never be underestimated as 'ease of use', as well as understanding, is a key in enabling large-scale change.

EXAMPLES OF BEST PRACTICES

- **Commitment to change**: EU countries, for example, are making significant progress and are an important example of best practice. However, some state governments in the US (for example, California) are also leading the way. The task is now to build on the success of these countries and states and establish targets from the post-Bali consultation process, to which all countries globally can commit themselves.
- **Taking direct action**: In 2009, the British Government announced a plan to provide cavity wall and roof insulation to all properties by 2015. 400,000 households a year will be fitted. A quarter of the UK's total CO_2 emissions come from homes. A further important policy decision was made by the British Government in 2009. A commitment is made that up to four coal-fired power stations would be fitted with carbon capture and storage (CCS) that takes the carbon dioxide from the burning coal and stores it underground. The real challenge is not in the technology. It is in creating the right environment for Government, Business and People to want to make the changes.
- **Combining the three agents of change**: Another best practice example comes from Australia, which has legislated to phase out non-energy saving light bulbs by 2012,[11] with a forecast 40 per cent energy saving as a result. This is a good example of the three agents of change – Government, Business and People, working cooperatively together to drive change. Business innovation and funding developed the technology, government has created the suitable legislative environment for change, and consumers support the initiative through their buying behaviour.

---

9   As reported by the Netherlands Environmental Assessment Agency. Website: http://www.pbl.nl/en/dossiers/index.html, CO_2 Climate Change, Frequently Asked Questions 10. Which are the top-20 CO2 or GHG emitting countries?

10   As reported on the Greenpeace website: http://www.greenpeace.org/australia/issues/deforestation/overview/forest-loss-climate-change. How forest loss causes Climate Change.

11   As reported on the BBC News website: http://news.bbc.co.uk/1/hi/world/asia-pacific/6378161.stm. Australia pulls plug on old bulbs.

The Role of Government, Business and People

GOVERNMENT

The contribution of government is important at all levels – international, national and state. The EU countries and an increasing number of American states are using a variety of approaches including regulation, legislation, tax and rewards.

- **Lead by example**: Although America is the world's largest producer of greenhouse gases, the dangers were consistently downplayed by the Bush administration. In a fundamental shift of policy the Obama administration, in April 2009, declared greenhouses gases were a threat to public health. The US Environmental Protection Agency (EPA) reported that US citizens, particularly the poor and those in ill health, were vulnerable to an increased risk of droughts and floods; sea level rises; more intense storms and heat waves; and harm to water supplies, agriculture and wildlife. It said that the science on man-made pollution as a cause of global warming is 'compelling and overwhelming'. Regulations will cover carbon dioxide emissions and five other greenhouse gases. This landmark decision puts the USA in a leadership position on climate warming.
- **Establish clear guidelines**: An important role of government is to set a clear statement of objectives, and the economic plans to achieve them. The following letter, written to the UN by a group of top business leaders prior to the Bali Conference in 2008, endorses the need for government action:

'It is our view, that a sufficiently ambitious, international and comprehensive, legally binding United Nations agreement to reduce greenhouse gas emissions will provide business with the certainty it needs to scale up global investment in low carbon technologies… We believe that tackling Climate Change is the pro-growth strategy.'[12]

- **Long-term planning**: Climate Change is already impacting regions and countries in a variety of ways and these will increase as temperatures continue to rise. It is important that responsible government not only works with other countries to agree greenhouse gas mitigation strategies, but also develops and deploys adaptation strategies to tackle the predicted impacts.
- **Collaboration**: In 2009, 75 countries signed a treaty to establish an international organisation – The International Renewable Energy Agency – to promote the use of solar power and other renewable energy sources. It will be a driving force behind renewable energy technologies such as wind, water and geothermal sources, demonstrating ways for countries to reduce their dependency on oil, gas and coal.
- **Legislation**: In 2009, the British Parliament passed the Climate Change Bill, the first of its type in the world. Government has committed to reducing carbon emissions by 50 per cent by 2050.
- **Establish a global financial framework**: There is already a best practice strategy operating increasingly successfully in areas such as the EU and California – the Cap and Trade system. This is a simple way of creating a predictable environment which

---

12 As reported at The Bali Communique on Climate Change: http://www.balicommunique.com/communique.html.

identifies clear limits to CO_2 emissions for businesses and individuals (Cap). Businesses that exceed their caps, can purchase 'offsets' to compensate, from businesses that overachieve (Trade). The importance of these offsets is that they can also be used to fund projects that support alternative energy production or efficiency such as forestry, wind, hydroelectric and solar power,[13] thus funding the innovation and technology development required to support the overall change.[14] In 2007, the carbon market was valued as $64 billion, with nearly $3 billion in carbon credits or 'offsets' traded.[15]

- **Reversing deforestation**: Another opportunity is reversing deforestation. An estimated 20 per cent of greenhouse gas emissions are from deforestation through the burning of forests with the resulting release of greenhouse gases. In parallel is the loss of the CO_2 absorbsion which occurs in forests. International and national government coordination to reduce this deforestation not only contributes to managing Climate Change but also addresses other challenges such as biodiversity and sustainability.
- **Fast-tracking reforestation**: A further task for government is in reforestation. There are already best practice programmes operating in China, Africa and Korea where reforestation programs are highly effective. However, the overall global forest loss versus reforestation is still negative. As funds become available from Carbon Offsets programmes there is the opportunity for many more wastelands to be converted back to forests.
- **Mitigation policies**: Effective mitigation is key in addressing Climate Change. The following table from the UN Environmental Programme outlines some selected policies and measures that can be used to mitigate Climate Change.

Nature	Policies	Measures
Target-orientated GHG emissions reduction measures	International	36 countries and the European Community accepted targets under the Kyoto Protocol
	State or Province	14 states in the United States, and many provinces in other countries adopted targets (Pew Centre on Global Climate Change 2007)
	City or local government	>650 local governments worldwide, and 212 US cities in 38 states adopted targets (Cities for Climate Protection – CCP)
	Private sector	For example, Climate Leaders Programme of USEPA – 48 companies (USEPA 2006)
Regulatory measures	Energy process and efficiency improvements	Energy efficiency portfolio standards, appliance efficiency standards, building codes, interconnection standards
	Renewable energy improvement	Renewable energy portfolio standards (RPS) Biofuels standard (for example, US Energy Policy Act of 2005 mandates 28.4 billion litres of biofuel/year in 2012) (DOE 2005)
	Raw material improvements	Industrial standards, research development and demonstration (RD&D)
	Fuel switching	Mandatory standards, RD&D
	Recycling and reuse	Mandatory standards, awareness creation, pollution tax

Figure 22   Selected policies and measures to mitigate climate change

---

13   As reported in *The Atlas of Climate Change* by Dow, K. and Downing, T. E. Comparative emissions of sample journeys, p. 75, Earthscan, 2006. Myriad Editions Ltd, www.MyriadEditions.com.
14   As reported in *How We Can Save the Planet*, Hillman, M. p. 118, Penguin Books, 80 Strand, London, WC2R 9RL, England.
15   As reported on www.copenhagenclimatecouncil.com/get.../tackling-emissions-growth.

Nature	Policies	Measures
Economic measures	Taxation policies	Carbon taxes, pollution tax, fuel taxes, public benefit funds
	Subsidy policies	Equipment subsidies for promotion of renewable energy sources
Technological measures	Technology commitments	Initiatives on strategic technologies, such as Generation IV Nuclear Partnership, Carbon Sequestration Leadership Forum, International Partnership for the Hydrogen Economy, Asia Pacific Partnership on Clean Development and Climate (USEIA 1999)
	New technology penetration	Technology standards Technology transfer, RD&D
	Carbon sequestration	Technology transfer, emission taxes
	Nuclear	Emission taxes, sociopolitical consensus
Others	Awareness raised	'Cool Biz' or 'Warm Biz' campaigns

Figure 22   Selected policies and measures to mitigate climate change
concluded

CLIMATE CHANGE IN CALIFORNIA

A Strategic Plan

The US Federal Government has for many years neglected the risks and challenges of Climate Change. President Obama in 2009 committed the administration to a new sense of urgency and a series of practical measures, acknowledging the grave nature of the problem.

However, cities and states have meanwhile been proactive. In 2007, the US Conference of Mayors launched the Mayor's Climate Protection Center to administer and track agreement by the cities to meet or beat the Kyoto Protocol in their own communities and to put pressure on Congress to pass bipartisan greenhouse reduction legislation. There are now 710 signatories to the agreement. Many states have made similar commitments.

California is of special interest. First because it produces 1.4 per cent of the world's and 6.2 per cent of total US greenhouse gases and therefore has the potential to make a major impact globally. Second, because under the leadership of Governor Arnold Schwarzenegger, the state has produced a long-term strategic plan, backed up by legislation and major education campaigns.

Schwarzenegger made it clear when he campaigned for Governor in 2003 that he wanted California to be number one in the fight against global warming – 'This is something we owe to our children and our grandchildren.'

Every aspect of the problem was studied in depth. For example, the risk of reduced water supply through more severe droughts, more winter flooding and the intrusion of saltwater into coastal aquifers. This could seriously damage agriculture, a $26 billion industry. Climate Change would increase levels of air pollution. Although California has developed one of the largest and most diverse renewable electricity generation industries in the world, less snow pack would reduce levels of hydrogeneration in the summer and fall. The risk of catastrophic forest fires would be increased. Changes in ocean conditions would adversely affect California's fishing industry. Some of these studies dealt with short-term problems but a long-term strategic view was also necessary.

> In 2006, California passed a landmark bill establishing a comprehensive programme of regulatory and market mechanisms to achieve real, quantifiable, cost effective reductions of greenhouse gases. Carbon emissions are to be reduced to 1990 levels by the year 2020, which is a 25 per cent reduction. Emissions will be reduced to 80 per cent below 1990 levels by 2050.
>
> The driving force for implementing this plan is the California Air Resources Board (CARB). An Environmental Justice Advisory Committee and an Economic and Technology Advancement Advisory Committee will advise CARB. These regulatory and legislative programmes have been reinforced by major communication campaigns and consultations so that everyone in the state is aware of the serious nature of the problems and the need for action.
>
> This is an inspirational case where political leadership has been translated into a coherent strategic plan backed up by legislation. The immediate challenge facing California is addressing the financial challenges arising from the global credit crunch whilst not weakening the implementation of the Climate Change plan.

BUSINESS

Why is it important that business takes an active part in addressing Climate Change? There are three key reasons:

- **Contribution**: In the developed countries, it is estimated that over 50 per cent of greenhouse gas emissions are generated by business activities. There is therefore a significant impact business can have in 'cleaning up its own backyard'.
- **Innovation and finances**: Achieving the changes required to reduce greenhouse gas emission reductions will require significant innovation and funding. Business clearly has a key role and self-interest in driving this. This is where the real benefit of the economic approach of Cap and Trade systems demonstrates its value. It not only sets the framework within which businesses know they need to operate, providing important predictability, but it also provides incentives for businesses to innovate and fund the development of energy efficient products and emission reduction.
- **Cost savings and efficiency**: Carbon reduction audit processes and strategies can also often result in more efficient use of resources, cost savings and efficiency gains for business. Managing Climate Change can not only be good for the planet but also for business. Johnson & Johnson achieved $30 million savings with a 3 per cent reduction in emissions. DuPont have achieved 67 per cent reductions in emissions since 2003 with a $2 billion saving since 1991. Eastman Kodak achieved 17 per cent reductions in emissions and $10 million saving by 2003.[16]

    Sir Terry Leahy, CEO of Tesco states, 'It is no longer an option for business to ignore Climate Change: it must be green to grow. Tesco has set itself challenging targets on

---

16   As reported in *The Atlas of Climate Change* by Dow, K. and Downing, T. E. Comparative emissions of sample journeys, p. 90, Earthscan, 2006. Myriad Editions Ltd, www.MyriadEditions.com.

carbon emissions, recycling and sustainable ways of working. We aim to lead by investing millions in technology and by helping our customers to make a big difference.'[17]

THE ARROW APPROACH

There are a range of options which can be used to address the challenge of Climate Change for business. These opportunities for moving towards non-carbon-based energy supplies can be best summarised using what can be called the 'ARROW' strategies. However, before these strategies are deployed, is essential to review the current status of the business with a Carbon Footprint Audit. This identifies strengths and weaknesses, opportunities and threats in relation to Climate Change and related issues for each business. From this the priorities for attention can be identified and the following approaches implemented using the ARROW framework:

1. **Avoid**: Avoidance of unnecessary activities which produce carbon emitting activities, for example, air travel.
2. **Reduce**: Based on the audit, specific plans for carbon emission reduction can be put in place. For instance, using low-energy light bulbs can reduce energy consumption by over 60 per cent. Simple notices reminding staff to 'turn off the lights' is another easy and low-cost initiative.
3. **Replace**: Where possible, carbon emission activities should be replaced. For example: replacing international face-to-face meetings with tele-meetings using technologies such as audio-conference, video-conference, teleconference and tele-presence facilities can not only reduce the carbon footprint, but can also significantly reduce costs and travel 'downtime'.

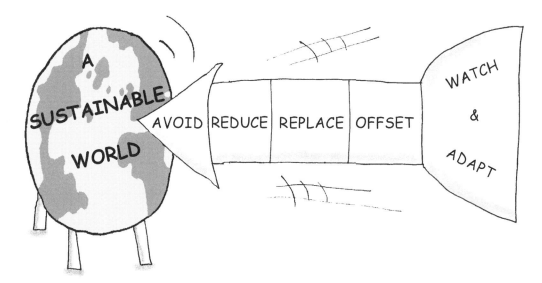

Figure 23   The ARROW approach to addressing Climate Change

---

17   As reported on http://www.independent.co.uk/news/business/news/tesco-follows-mamps-with-climate-change-move.

4. **Offset**: Where it is not possible to reduce or replace, then offsets are available to counter the impact of greenhouse emissions now available through the rapid growth of carbon trading. In using offsets there are two important considerations. Firstly, offsets should be used only where other alternatives are not available. They should be seen as a 'last resort'. Equally important is ensuring the quality of the offsets projects by using only those which are properly audited and have standards which have a demonstrated positive impact on reduction of carbon emissions.
5. **Watch and adjust:** The final strategy is to set targets and monitor and adjust them to ensure that the overall strategy is being successfully implemented. Continuous improvement is essential.

A GREENER WORLD USING ICT

Information and Communications Technology (ICT) companies can play an important role in the reduction in the global carbon footprint – both directly, and through the services they provide to other companies. Currently representing 2 per cent of the global carbon footprint, ICT services are continuing to expand. However, there are important changes in work practices that many ICT companies are taking to reduce their company-specific carbon footprint – such as collaborative communications to reduce travel, data centre and facility consolidation/virtualisation, and new techniques for power and air conditioning.

Further to this, effective use of ICT services presents an even greater opportunity for all companies to reduce their carbon footprint, reduce costs, and achieve their corporate and environmental sustainability objectives. Total carbon reduction opportunities presented by effective use of ICT services could represent more than 10 per cent of the global carbon footprint.

Orange, operating a range of global telecoms services including mobile, global data and voice services, is a good example of the types of services provided which can be used to reduce carbon footprints – both internally, and with customers and suppliers. Orange has a strong focus on reducing its internal carbon footprint, and was recently identified as the leader in Green IT by Verdantix in it's Green Quadrant for Sustainable Telecoms in Europe. Orange also has a comprehensive programme of services which enable any business to reduce their global carbon footprint. Examples of approaches that can be adopted by business are:

- smart air and ground travel management by using collaboration communications tools such as telepresence and videoconferencing;
- work at home policies and communication tools, which can significantly reduce employee travel times;
- paper-free workflows are another key area where for examples of order and delivery processes are completely driven by IT infrastructure;
- ITC providing real-time logistics and communications technology to manage everything from just in time ordering and delivery to vehicle fleet management. Tele-meeting and M2M (Machine to Machine) technologies to improve the efficiency of the supply chain;
- data centre consolidation and IT virtualisation to maximise the utilisation rate of servers and therefore dramatically reduce energy consumption.

According to Helmut Reisinger, SVP of Orange Business Services, 'There is an important opportunity to leverage ICT to enable green business practices throughout a company and use it to drive business innovation. As many of these innovative and supply changing management practices reduce cost as well as carbon footprint, large and small companies are increasingly turning to Information and Communications Technology to deliver on both their cost and sustainability objectives.'

PEOPLE

In the UK, for example, household usage contributes to an estimated 30 per cent of national energy consumption. It is therefore important to look at the strategies that people, particularly in the top 20 countries, can use to reduce their personal carbon footprints. The table below outlines average UK household usage.

Table 1   Direct greenhouse gas emissions of the typical UK individual[a]

		Tons
House	Heating	1.2 (20%)
	Water heating	0.3 (6%)
	Cooking	0.1* (2%)
	Lighting	0.1 (2%)
	Electric appliances	0.6 (10%)
Total household		**2.3** (38%)
Car		1.2 (20%)
Bus, rail		0.1 (2%)
Air travel		1.8 (30%)
Total Transport		**3.1** (42%)
House and transport		**5.4** (90%)
Other direct emissions		0.6 (10%)
TOTAL		**6.0** (100%)

[a]   As reported in *How to Live a Low Carbon Life*, Goodall, C., Earthscan, 8 Camden High St, London, MW1 0JH, 2007.
*   Assumes that cooking is done by gas.

Focus

A quick review of the following table shows that a major difference can be made by focusing on the key areas which can make the most impact – electricity supplies, air travel and car transportation.[18] Addressing these three key areas could contribute to a target reduction of over 50 per cent over the next 10 years in a typical UK household. If every household in the

---

[18]   As reported in *National Geographic Magazine*, Special report 2008: Changing Climate, Village Green article, Nijhuis, M. p. 72.

54   Developing a Plan for the Planet

UK undertook this challenge, the impact on greenhouse gas emissions in the UK could be quite significant. So let's look at some ways this could be achieved.

- **Renewable energy supplier**: Immediately switching to a renewable electricity supplier would reduce CO_2 emissions by one ton, an immediate 17 per cent reduction in greenhouse gas emissions. This is well on the way to a 25 per cent reduction targeted for the first five years – and a 50 per cent reduction by 2020.
- **Reduce or offset air travel**: Reduction of air travel or use of offsets for this travel would effectively reduce effective emissions by a further 1.8 tons (30 per cent).
- **Reduce car emissions**: Replacing current vehicles over the next 10 years with smaller engine size or hybrid vehicles could reduce total emissions by a further 5 per cent. Some energy and transport companies are offering offset programmes which, if used, could result in the net carbon emissions reducing to zero.

These three strategies alone, some of which can be implemented immediately, and some of which can be phased in over time, would bring the total effective emissions reduction to over 50 per cent by 2020 per household – well above the EU target of 20 per cent in a similar period.

- **Further reductions**: These can be achieved now, or phased in through approaches such as updating a heater or boiler systems to increase efficiency, improving insulation, telecommuting (working from home on a regular basis) and increased use of public transport. These further reductions could quickly reduce emissions to above the EU 80 per cent objective.
- **Solar-powered clothes drying**: It has been estimated that in the US, a household's carbon footprint could be reduced by 700 pounds by 'air drying' clothes 6 months of the year. This is particularly applicable in the warmer regions, and is highly utilised in countries such as Australia where powered clothes dryers are used much less often. It has been estimated that if everyone in the US was to use air drying instead of electric drying, it could reduce the US carbon footprint by over 10 per cent.

Figure 24   You know these solar-powered clothes dryers work really well, and are very simple to use!

- **Greenhouse reabsorption:** Other ways people can play an active part is by community programmes such as reforestation programmes/community tree planting activities, reclaiming 'non-productive' land and nurturing green zones. The CimateXChange programme in Oxford, UK is an excellent example of these types of programmes and their ability to engage the local community in proactive action (see box insert).

> **CLIMATEXCHANGE – AN OXFORDSHIRE PARTNERSHIP**
>
> When groups of concerned people in a community join together they are able to make a significant difference.
>
> Take the example of the ClimateX project in Oxfordshire, a UK county with a population of over 600,000 people.
>
> Led by the Environmental Change Institute at Oxford University, ClimateXChange seeks to create 'climate buzz' by enabling everyone – individuals, families, neighbours, pubs, schools, businesses, churches – to 'get fantastically imaginative about Climate Change'. Partners include schools, local councils, faith organisations and many more. ClimateX helped with over 150 events in its first year and 50 local Climate Change groups now exist.
>
> Meeting people 'where they are' and inspiring them to take their next step has been the cornerstone of ClimateX. Resources and activities range from thermal imaging cameras to low carbon scarecrows, including film screenings, Question and Answer talks, hands-on 'Eco-homes Open Days', surveys, competitions, posters, poetry and creative responses. Sharing best practice, celebrating others' projects and encouraging peer-to-peer learning (joining the dots) has been crucial. Volunteer 'climate explorers' in the community advise and help, while out of these grassroot efforts there is now growing engagement with the county's leadership.
>
> This a wonderful role model of a community working together to deal with Climate Change issues. More information is available on the web site: www.climateX.org.

CHAPTER 4

Executive Brief No. 3: Energy Supplies

'Energy is like any other economic good. It needs decent governance, functioning institutions and effective markets to get electrons from the producer to the consumer on a sustainable basis. Without reliable energy, virtually every aspect of life is negatively affected. After all, energy, at its most basic, is the capacity to do work.'

Thomas L. Friedman

Figure 25 **We really should have checked the oil before we put our foot on the accelerator!**

SUMMARY

The world's economy is driven by a few key energy supplies with oil, coal and gas being the main sources supporting the increasing population on Planet Earth. Each is a major source of greenhouse gases. Increased consumption therefore worsens the Climate Change problem. In addition, oil, a major source of energy for transport and agriculture, has arguably already reached peak production.

The current priority is to rapidly develop and deploy sustainable sources of energy and move to non-carbon-based energy economy. In essence this will require what many are already calling a 'new green revolution'.

58 Developing a Plan for the Planet

> This can be achieved through targeting increased efficiencies in the use of energy resources and the transition to long-term sustainable sources, which also reduce greenhouse gas emissions.
>
> Government, Business and People can all help with this, through the use of the ARROW Strategies – Avoid, Reduce, Replace, Offset and Watch.
>
> Long term, this transformation will provide the foundations for a sustainable world.

The Current Situation

- **Energy use increasing**: Global energy usage has been increasing by approximately 2 per cent per year.[1] The result is that by 2008, 50 per cent more oil, gas and coal was used than in 1980 and emissions from fossil fuels were 30 per cent higher than in 1990 – the baseline for the Kyoto Targets.[2]
- **Non-renewable sources**: The vast majority of this energy comes from non-renewable resources such as gas, coal and oil. Most significantly, they create emissions that contribute to global Climate Change and local air pollution problems.

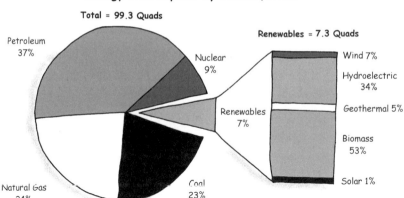

Figure 26   US Energy 2008

Source: As reported by the Centre of Sustainable Energy, http://css.snre.umich.edu/css_doc/CSS03-12.pdf.
Data Source: U.S. D.O.E Energy Information Administration (EIA) (2009) Monthly Energy Review 2008.

- **Peak oil**: It has been argued that easily accessible oil supplies have reached, or are reaching, their peak with production therefore forecast to decline.[3] This will have a major impact on transport and agriculture, which are highly dependent at the moment on oil for production and operation. The International Energy Agency reported in 2009 that more than 800 oilfields in the world, covering three-quarters of global reserves, have already peaked. There has been chronic under-investment by oil-producing countries. As the global economy returns to growth, demand will

---

1   As reported on http://www.wri.org/publication/content/8601.
2   As reported on http://flatplanet.wikispaces.com/Group+11+Fossil+Fuels+and+Sustainable+Energy.
3   As reported on http://www.relocalize.net/is_the_iea_admitting_the_peak_has_been_reached.

far outstrip supply – resulting in higher prices not only of energy supplies, but also for other resources such as food, which are highly dependent on oil supplies for production and transportation.

- **Energy poverty**: This is also a major problem. At present, over 1.5 billion people, about one-third of the developing world's population, are without access to modern energy and are therefore forced to rely on carbon-emitting biomass and fossil fuel energy,[4] further contributing to deforestation and greenhouse gas emissions. Without reliable energy many aspect of life are damaged. Cooking over open fires creates air pollution and greenhouse gases. Without electricity, computers are not available so there is no access to the knowledge in the World Wide Web and no mass communication. Drugs cannot be refrigerated. Unreliable supplies damage productivity. For example, in 2007, only 19 out of 49 power plants were working in Nigeria. Nepal currently only has electricity available for four hours per day.
- **Business impacts**: Some parts of the global economy will be more affected than others by the declining oil supplies simply because they are more oil intense. The automobile industry, food production, as highlighted, and the airline industry are particularly susceptible.
- **Alternative energy sources**: The challenge is to raise production of non-carbon-based energy sources before the demand for oil exceeds the available supply. Despite this imminent possibility, alternative non-carbon-based sources currently supply less than 10 per cent of the global energy requirements of the human population.

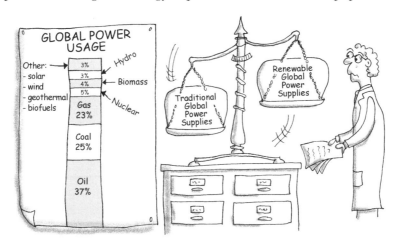

Figure 27   I think we need to do some more work on the renewables!

- **Cross impacts of some alternative energy sources also need to be considered**: Some alternative energy sources, such as bio-fuels, are driving up the prices of grain and other food sources as land is increasingly devoted to its production. A further side effect is that significant areas of land are deforested at a time when forests are increasing in importance due to their CO_2 absorbsion capabilities. Other alternatives, which are not carbon-based energy sources, include nuclear fuels, but these have waste storage and safety issues which are yet to be fully resolved.

---

4   As reported on the World Bank website: http://web.worldbank.org/WBSITE/EXTERNAL/NEWS/0,,content MDK:20127296~menuPK:34480~pagePK:34370~theSitePK:4607,00.html.

Opportunities and Best Practice

ALTERNATIVE ENERGY DEVELOPMENTS

- **Solar power**: Solar power is already a viable non-carbon-based energy supply, however, the real challenge for solar power is the speed of investment and development. Best practices, however, are already emerging. Faced with a doubling of the present UAE population of 4.5 million people in the next 50 years, the UAE has made a $15 billion commitment to renewable energy, including some 2,000 megawatts of concentrated solar power. This also has the strategic advantage of preserving reserves of oil.

Figure 28 I can't seem to get these numbers to add up!

- **Water**: Energy from water sources such as hydro-electricity plants is already a significant source of energy in many countries. Large-scale hydro-electricity, however, often involves large reservoirs which result in formation and release of methane from decaying biomass. Small-scale hydro-electricity facilities avoid these problems.
- **Clean coal**: Coal is still in abundant supply and technology is being developed to potentially remove the greenhouse gases emissions before they leave the power plant. For example, two coal-fired plants planned in Kansas, US, have planning permission on the condition that they build one plant using new clean technology and develop wind energy on the side. Many countries using coal for energy have demonstration plants and incentives to convert to clean technology. China has emerged in the past two years as the world's leading builder of more efficient, less polluting coal power plants, mastering the technology and driving down costs.
- **Wind**: Wind turbines of a variety of sizes are being used to generate electricity for both the national power grid, for isolated communities and for homes. In 2009, the London Array wind farm was authorised with a UK Government subsidy. The first phase will have 175 turbines with a capacity of 630 megawatts. This wind farm will be four times the size of the largest offshore wind farm in operation today.
- **Tidal, wave and ocean**: The movement of the sea is being used to generate electricity and is increasingly being harnessed in countries such as the UK.

- **Geothermal**: The earth's natural heat is being increasingly developed to fuel power plants. Smaller-scale operations are also being harnessed to use the constant temperature 1.5 metres below the earth's surface to heat and cool buildings.[5]
- **Biomass**: Purpose grown or waste plant material can be used to generate electricity and heat which is carbon neutral so long as the CO_2 released is the same as the amount removed.
- **Smart grids**: US President Obama has outlined plans to create a 'smart grid' that uses information technology to manage flows of power. It would be linked with wind farms in South Dakota and solar arrays in New Mexico. The cost would be $1,000 billion, however, with current spending of $400 billion a year on electricity, a 10 per cent reduction in power use would recover costs in 25 years.
- **Nuclear fusion and nuclear fission**: There has been much debate about whether nuclear power is a long-term viable alternative source of energy due to the very real concerns about waste and security of the most common form of nuclear power, nuclear fusion. The advantage of this source of non-carbon-based fuel is that it is available now and being used extensively in countries such as France, and provides a ready supply of non-carbon-based energy – although dealing with the waste and security issues still need to be properly addressed. The other potential form of nuclear energy, nuclear fission, provides a highly attractive longer-term option, but it still requires further work to bring it to a practical and commercially viable level. However, it has the advantage of not having the waste and security issues as the by-product is water.
- **Efficiencies**: The human population, particularly in the developed world, has had low-cost and abundant energy supplies for over 100 years, resulting in high levels of inefficiencies developing. The opportunity therefore exists to significantly reduce energy usage simply through more efficient behaviours, through both the development and use of much more efficient appliances and energy conservation technologies, as well as changes in business and consumer usage patterns.

These alternative supplies need full industrialisation and expansion to ensure commercial viability on the scale required to replace carbon-based fuels. The rapid rate of developments in many of these technologies means that this is potentially feasible. However, it will require a significant acceleration of investment and incentives. Countries such as Germany, which are leveraging these investments and incentives, are quickly becoming leaders in these technologies and therefore are increasingly well positioned to lead in the new green revolution.

OPPORTUNITIES

Focus

As already highlighted, a major impact can be made by focused efforts to increase energy efficiencies and use of sustainable energy supplies in the major energy consuming countries. This is particularly evident when we review the BP 'Statistical Review of World Energy 2008' which shows that nearly half of global oil consumption in 2007 was in the top six energy consumption countries.

---

[5] Reproduced with permissions from *The Atlas of Climate Change* by Dow, K. and Downing, T. E, Comparative emissions of sample journeys, p. 83, Earthscan, 2006, Myriad Editions Ltd, www.MyriadEditions.com.

Table 2   Oil consumption for six key countries

Country	Percentage	% Subtotal
USA	23.9%	
China	9.3%	
Japan	5.8%	39% for the top 3 countries
India	3.3%	
Russia	3.2%	
Germany	2.3%	47.8% for the top 6 counties

A focused approach on the top energy consumption countries, in the areas of transport, industry and services could therefore significantly impact the energy consumption on Planet Earth.

The Role of Government, Business and People

GOVERNMENT

- **Drive for energy efficiency**: By introducing simple ways of increasing energy efficiencies, as well as providing incentives for use of sustainable energy supplies, a dramatic impact on the reduction of carbon-based energy consumption and increases in use of sustainable energy supplies can be achieved.
    - **Cars and motor transport**: Best practice improvements are already being achieved in fuel efficiency in cars and motor vehicles. For example, a study published by the Society of Automotive Engineers highlighted that advanced turbo-charged engines could use 40 per cent less fuel than conventional engines. Governments in developed countries can introduce stricter standards such as Corporate Average Fuel Efficiency (CAFE) to help 'push and pull' the technology into the marketplace and make sure it is used effectively.[6]
    - **Household appliances**: Introduction of consistent efficiency labelling on all consumer products can give much greater consumer choice of products and the opportunity to reduce energy consumption in the normal replacement cycle. Application of simple technology changes by industry such as the use of On/Off switches rather than standby only can also reduce energy consumption.
    - **Building construction**: In the UK, 20 per cent of the average energy usage in a home is spent on heating. This can be significantly reduced through the introduction of better insulation, such as installation of wall cavity insulation in initial construction of housing and in renovation. Government regulation on building standards and incentives to introduce these changes to existing homes

---
6   As reported in *Earth: The Sequel*, Krupp, F. and Horn, M., p. 226, W. W. Norton & Co. Inc, 500 Fifth Avenue, New York, NY. 10110. Website: www.wwnorton.com.

can play an important part in achieving reductions in energy usage. The UK is already taking active steps to accelerate the adoption of these practices.
- **Research and development**: Lack of consistent guidelines and incentives for sustainable technologies in countries such as the US over the last decade have led to extremely limited investment in these technologies. Other countries such as Germany, which have demonstrated a consistent and long-term view on developing sustainable energy technologies, have had much more significant investment, development and usage of sustainable technologies.

    The opportunity therefore exists for a massive increase in investments for new energy sources driven by clear long-term government policy and incentives. This is critical, Thomas L. Friedman, in his book *Hot, Flat and Crowded,* states that in 2007 R&D investment in US electric utilities was only 0.15 per cent of total revenues. Many comparative industries have 5–10 per cent investment in R&D. Clearly, increased investment and development will significantly accelerate the commercial viability of sustainable energy technologies.
- **Agriculture**: 70 per cent of energy used by the agriculture industry is for water transport. Yet, there are many much more highly energy-efficient ways of transporting water than are currently employed. Introduction of 'water purchase and trade' systems, such as those introduced in Australia, are already successfully driving much more efficient use of water through distribution efficiencies and introduction of more water efficient crops.
- **Aviation**: Air travel is a part of modern life, yet it is the highest energy consumption and greenhouse gas-emitting form of travel per mile. Air miles emit twice as much CO_2 per mile as a car and more than ten times that of a train or coach. Government can play an important part by introducing to the airline industry the same fuel tax system that is paid by other transport suppliers – bringing the real cost of air travel in line with other transport industries. The introduction of Cap and Trade for all CO_2 emissions can ensure that the airline industry contributes proportionally to the cost of the carbon it generates.
- **Refocus subsidies**: Many governments of the top energy-consuming countries are still substantially subsidising the oil and coal industries. This is a legacy from the last century when governments were lobbied to support the oil and coal industries. These subsidies are significant: in the EU, direct and indirect subsidies to the coal industry in 2001 amounted to 13 billion euros and 8.7 billion to the oil and gas industry. In some of the top energy-consuming countries, these subsidies have actually increased. In 2005 in the US, a further $2.5 billion was given to the coal industry and a further 1.5 billion to oil and gas firms.[7] Governments have the opportunity to establish a policy of steadily removing oil and gas subsidies. This approach, together with the Cap and Trade strategy, could provide the opportunity for incentives and funding for the much more rapid development and deployment of sustainable energy supplies.
- **Drive innovation**: Cap and Trade provides countries with a predictable economic and business environment that businesses need to develop and bring to market the alternative technologies required to develop more sustainable energy strategies. The benefit of interconnectivity is that the Cap and Trade strategy also drives the move to more sustainable energy supplies.

---

[7] As reported in Brown, L., *Plan B 3.0*, p. 55, Earth Policy Institute, W W Norton & Company. 500 Fifth Avenue, N.Y. 10110.

BUSINESS

Each business has the opportunity to develop strategies for energy sustainability. This, as we have seen, is not only good for the environment, but also due to the significant cost savings it can result in, makes good business sense. Again, these can be based on the ARROW approach:

1. **Avoid**: Where possible, avoidance of unnecessary activities that use energy is an important strategy. A simple example is using audio and video conferencing to reduce the need to travel to business meetings.
2. **Reduce**: The obvious areas are reduction in the high-energy consumption areas such as lighting, refrigeration and facilities requirements. Technology solutions exist where lighting is automatically switched off when the room is not in use, through movement and heat sensors. Enclosing refrigeration facilities, for example in supermarkets, significantly drives down the energy consumption requirements as well as the costs. Homeworkers practices need less office facilities and use less energy. The time and energy usage of employees getting to and from a fixed workplace is also reduced. SKY television remotely reduces power useage in household units, when not in use.
3. **Replace**: Where possible, energy consumption activities can be replaced with alternative cost and energy-efficient technologies. This can be as obvious as using low-energy light bulbs and the introduction of more energy efficient equipment – everything from computers to transport. Business can use sustainable energy supplies or suppliers. Are there sufficient sustainable energy resources to support this change? Not at the moment. However, if the demand for these sustainable energy supplies is there, this will provide the incentive for companies to grasp the commercial opportunities and, through innovation, deliver the volume and quality required.

    Use of alternative technologies and manufacturing can similarly have a significant impact on reduction of energy and CO_2 levels. A good example of the development of best practice in this area is in the cement industry. Production of cement, a key component of the building industry, is a contributor to both high greenhouse gas emissions and high energy consumption. However, this can be significantly reduced by moving to alternative cement production techniques such as polymer cement.
4. **Offset**: Where it is not possible to reduce or replace, then offsets are available to counter the impact of non-sustainable energy consumption. Increasingly, airlines are introducing the option for passengers to select offsets when purchasing air transport tickets.
5. **Watch and adapt**: as with any business strategy, establishing clear and measurable targets and monitoring them to ensure that the overall sustainable energy strategy is being implemented is key in achieving demonstateable improvements in these areas.

    Increasingly, companies are recognising that being part of the 'new green revolution' is no longer an option – it is an increasingly important part of a long-term sustainable business strategy.

    Consumers and customers are now regularly demanding demonstrable sustainability strategies from their suppliers – otherwise they are excluded from the bid and supply process.

    The ability to not only implement, but also to monitor, adapt and improve sustainability strategy in business is therefore increasingly becoming a must have. The development and importance of the role of Chief Sustainability Officer (CSO), often reporting directly to the CEO, is a clear demonstration of the importance many companies are placing in this area.

PEOPLE

Approximately thirty per cent of energy consumption is domestic and therefore people have a key role to play in reducing energy consumption and driving the increased use of sustainable energy supplies. Often consumers do not know where to start in achieving this. However, there are many online household energy and CO_2 auditing tools available, as well as service organisations which will provide energy consumption assessments and monitor progress. A review of existing and ongoing energy usage, however it is done, quickly demonstrates where significant changes can be made – and progress in achieving these changes – making it increasingly easy for people to play an important part in driving change in this area. Using the chart below, we can quickly see how quickly a household can implement changes which significantly reduce non-sustainable energy consumption.

Table 3   UK Energy consumption estimates[a]

	Energy Type	Percentage
1	Heating & Cooling (including water heating)	25%
2	Lighting & Cooking	15%
3	Car & Ground Travel	20%
4	Air Travel	33%
5	Others	7%
	Total	**100%**

[a] Authors estimates from various sources including: www.carbonfootprint.com/energyconsumption.html.

From this table, it can be seen that approximately 40 per cent of individuals' energy consumption in the UK is for household usage. A further 58 per cent is for transport, so these are the key areas to tackle in strategies for addressing sustainable energy consumption. It is also useful consider the typical energy consumption of appliances within the household:

Table 4   Typical energy consumption of appliances within the UK household

Electrical equipment	Electricity consumption (terawatt hours per year)
Consumer electronics (TVs, computers, phones, etc.)	10.4
Washing machines, dryers and dishwashers	11.8
Cookers, kettles and microwaves	11.9
Lights	17.4
Fridges and freezers	17.5
Total	**73.0**
In 2004 power stations in the UK emitted 47 million tons of carbon. The figures are for the nation as a whole.	

Source: Environmental Change Institute, Oxford University.[a]

[a] As reported in *Heat: How to Stop the Planet Burning*, Monbiot, G., p. 7, Allen Lane Publishers, 28 September, 2006.

Over one-third of energy consumption is taken up by lighting and refrigeration, with washing, cooking and electronics comprising the second key areas to focus on. Again, the ARROW approach can be used in addressing household usage.

1. **Avoid**: Developing realistic strategies to avoid household and transport energy usage is a complex task as our personal and business lifestyles are often based on the efficiencies provided by the 'labour-saving' appliances and transport means.
2. **Reduce**: Smart thinking about energy consumption can significantly reduce not only the energy usage but also the household cost of these services.
3. **Replace**: Replacing an existing non-sustainable energy supplier with a sustainable energy supplier can reduce CO_2 emissions as well as driving development and deployment of new technologies. For example, wind power installations increased globally by about 29 per cent in 2008, with the US leading the way. Moving from air to rail or bus transport can also make a significant impact as seen in the chart opposite.

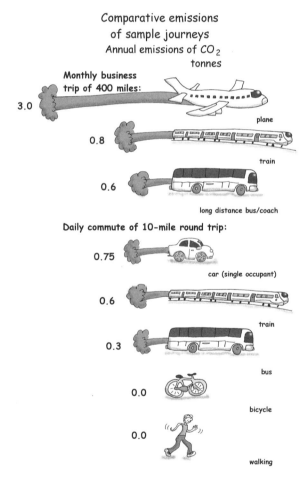

Figure 29   Comparative emissions of sample journeys

Source: Reproduced from *The Atlas of Climate Change* by Kirstin Dow and Thomas E. Downing, "Comparative emissions of sample journeys", p. 89, Earthscan 2006. Copyright 2006 Myriad Editions Ltd, www.MyriadEditions.com.

4. **Offset**: Energy can be offset using the same strategies as outlined in our review of Climate Change. Purchasing of offsets is important, as the offsets fund not only greenhouse gas emission reduction but also the development of sustainable energy projects.
5. **Watch and adapt**: Smart thinking about energy use involves setting up a plan on how to best manage household energy consumption, checking how well objectives are being achieved and then adapting to improve performance

WORKING TOGETHER TO INCREASE RESOURCE EFFICIENCIES IN SCHOOLS

A very good example of 'Smart Thinking about Energy' is taking place across an increasing number of schools and local councils in the UK, using technology developed by a local company, Optimal Communications. This equipment monitors and provides a report on a schools' resource usage including electricity and water. A leading UK council has teamed up with a number of schools in the area to introduce both energy saving, and green technologies into both the schools' day-to-day operations and curriculum.

Students are able to monitor the school's usage of key resources and address issues of both efficiencies and behaviour. The monitoring programme is run by members of the school. A shining example of this is where 11-year-old children are responsible for monitoring and addressing the school's energy needs.

This is an ideal example of where partnerships between business, government and people can work together to begin successfully tackling the Global Challenges.

THE PROBLEM OF POWER

Since the dawn of civilisation, with the discovery that fire could be used to cook food and provide heat, and that animals could be used for transport and agriculture – the human race has become increasingly dependent on all forms of power. Modern civilisation is now so totally dependent on the supply of energy and power to function, that were these power supplies to cease, much of human civilisation would come to an abrupt halt.

With more than half the world's population now living in cities, this dependency is even more important since the breakdown of even the electricity grid would have a devastating effect on all the basic elements of society. The movie *Die Hard 4.0* highlights the impact of this. Electricity supplies are suddenly cut to a large region, immediately causing chaos in communication, transport, health and security systems.

Telephone systems would be unusable, and even mobile phones would soon run out of battery life. Communication systems would quickly breakdown.

Money would immediately be a problem as accessing cash through ATMs would no longer be possible. The global financial systems, which are completely dependent on computer systems for the transfers of money, would quickly come to a halt.

Transport would quickly be affected as traffic management systems break down and even filling cars with petrol at a service station would not be possible, as the electricity would not be available to power the pumps. This would quickly impact transport and logistics, and therefore the supply of essential services would also be quickly affected.

Availability of food resources would also be quickly impacted, not only due to the problems of transport. Refrigeration would no longer be available to store food – which would particularly affect the hotter regions of Planet Earth.

As the supply of these services break down there would be an increased requirement for effective security and control. However, the security and policing systems would also be weakened by the breakdown of communications, transport and information systems.

Communication systems would break down as television, radio services and Internet services could no longer operate. This would seriously affect the ability to keep people informed and to manage the crisis. Health systems would be impacted as refrigeration would no longer be available to store vital drugs and electricity no longer available to supply life-support systems.

How different this picture is to even 100 years ago, where a breakdown in the electricity supply would not have greatly impacted people and the community. If we go back even 50 years, there were few computer-driven systems. The damage, though significant, would have been manageable.

In the twenty-first century, our complete dependency on reliable and consistent supplies of power to operate our modern civilisation is absolute. Yet as our dependency increases daily, the traditional sources of supply, such as oil, are declining or causing Climate Change damage

It is important, therefore, that in looking at a Plan for the Planet that we spend time understanding the limited sources of historic power end energy, and how they are being affected by the changes currently taking place on Planet Earth, and how we can fast track the development of alternatives.

CHAPTER 5

Executive Brief No. 4: Water and Food Supplies

'All too often, water is treated as an infinite free good. Yet even where supplies are sufficient or plentiful, they are at increasing risk from pollution and rising demand.'

Kofi Annan, Former Secretary-General UN

Figure 30   The world's aquifers are running low!

SUMMARY

The trebling of the world's population since the 1950s has increased the demands on the world's water and food supplies. This is particularly evident in South East Asia and many states in Africa, where the rapid population growth is outstripping the demand on land and food production resources, moving these countries from net food production countries to food-importing countries. Modern technologies have increased the ability of the human race to keep increasing food production. However, this is putting additional stress on the available water supplies since 70 per cent of water is needed for the production of food.

Solutions for water and food must be focused not only on effective production and utilisation but also appropriate storage and distribution capabilities to enable equitable distribution of food supplies to all regions.

> In parallel is the need to manage declining non-renewable water supplies – such as declining aquifer volumes and lake, river and glacier resources. These are all key sources of water for human consumption and food production. Improved productivity and creative conservation strategies are essential in achieving sustainability.

The Current Situation – Water Supplies

- **Limited supplies**: Less than 3 per cent of the world's water supply is fresh water. The rest is either saltwater or undrinkable. Of this 3 per cent, 2.5 per cent is locked up in the ice of Antarctica, the Arctic and glaciers, and therefore not immediately available to for human usage. The human race therefore relies on the remaining 0.5 per cent for its entire ecosystem and its fresh water needs.[1] A recent report by the UN World Water Assessment Programme states 'that urgent action is needed if we are to avoid a global water crisis'.

    Globally there is currently no overall shortage of fresh water – one of the key challenges is, however, ensuring that it is available where required, particularly in regions where rapid population growth is outstripping the available supplies. More than 1 billion people already live in areas where water is scarce. By 2030, the UN predicts that 40 per cent of the world's population – over 3 billion people – will be living in 'water scarcity' regions.[2]

    The other key issue is inefficient water use and water wastage. In most European cities, with 100,000 people or more, ground water is being used faster than it can be replenished. It is also estimated that 40 per cent of water in European cities is being wasted.[3]
- **Current usage**: On a global scale, it is estimated that approximately 10 per cent of the fresh water supply is used for domestic purposes, 70 per cent for agricultural and 20 per cent for industrial, of which the largest industrial use is for cooling in thermal power generation.[4] Water therefore is plays a critical role in both food and energy production.

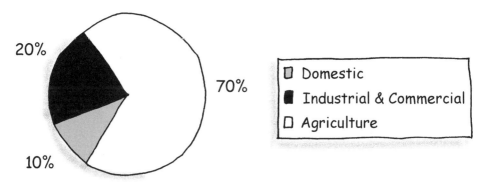

Figure 31   Global fresh water usage estimates

---

1   As reported in *The Atlas of Water*, Clarke, R. and King, J. p. 20, produced for Earthscan by Myriad Editions Ltd., 59 Lansdown Place, Brighton, BN3 1FL, UK. Website: http://www.MyriadEditions.com.
2   As reported on the UNFPA website: www.unfpa.org/swp/1999/newsfeature1.htm.
3   As reported on the Globnet website: www.globe-net.com/news/print_pdf.cfm?type=2&newsID=3467.
4   Authors estimates from various sources.

- **Increasing demands**: Global economic growth, population pressures and the rise of mega cities have all driven water usage to record levels. Today, around 3,800 cubic kilometres of fresh water is withdrawn annually from the worlds lakes, rivers and aquifers, which is twice the volume extracted 50 years ago.[5]

KEY IMPACTS

- **Depleting underground aquifers**: One of the critical issues is that non-renewable underground aquifers, which have been supplementing above-ground water supplies for thousands of years in many regions, are running low. This is taking place across the planet.
    - **China**: The deep non-renewable source of water which supplies much of the water for the Chinese grain harvest is dropping at an estimated three metres or ten feet per year – yet it is estimated that two-thirds of China's water supplies are used for growing food on inefficient farms.
    - **USA**: Wells have gone dry on thousands of farms in the great Southern Plains as a result of a drop in the underground water supply by over 100 feet or 30 metres.[6] This issue is critical, as this region is the breadbasket of America – supplying about one-fifth of the total American agricultural harvest. Rapid depletion of non-renewable water supplies in this region clearly impacts the ability of the region to continue to provide this key source of food supplies.
- **Impact on Planet Earth's lake systems**: In addition to the reduction in underground aquifers, many lakes are rapidly decreasing in size or drying up completely and thus unable to support the increasing populations that rely on them. Again, this is taking place in regions across the planet as population expansion continues to increase demand on supplies.
    - **China**: In one of the main regions through which the Yellow River flows, Quihai province, the number of lakes has halved in the last 20 years from over 4,000 to less than 2,000 lakes. Around Beijing, with high demands from population, industry and agriculture, the number of lakes has reduced by over 90 per cent, with less than 100 of the original 1,000 lakes remaining in the area.[7]
    - **Israel**: The Sea of Galilee and the Dead Sea are both shrinking fast as the water from the Jordan River is used before it can supply these lake systems.[8]
    - **Nigeria, Niger, Chad, Cameroon**: Lake Chad has shrunk to 10 per cent of its former size through persistent drought[9] and increased consumption driven by the rapidly rising population in these regions.
    - **USA**: Owen Lake, in California, which covered 200 square miles in 1900, has disappeared completely following the diversion of the Owen River to supply the water requirements of Los Angeles.[10]

---

5 As reported in *The Atlas of Water*, Clarke, R. and King, J. page y, Produced for Earthscan by Myriad Editions Ltd., 59 Landsdown Place, Brighton, BN3 1FL, UK. Website: http://www.MyriadEditions.com.
6 US dept. indicators.
7 As reported in *Peoples Daily*, 20 natural lakes disappear each year in China, Heng, L., 21 October, 2002.
8 As reported on BBC, Dead Sea to disappear by 2050, Hawley, C., 3 August, 2001.
9 As reported on UNEP website: www.na.unep.net.
10 As reported on http://www.american.edu/ted/mono.htm.

Figure 32   I now pronounce the Dead Sea officially dead!

- **Russia**: The level of the Aral Sea, once the world's fourth largest inland sea, fell by 16 metres between 1962 and 1994. The water volume has dropped by about 80 per cent due to extensive irrigation since the two rivers feeding the sea were diverted in a Soviet scheme to grow cotton in the desert.[11]

Figure 33   I can remember when the fish used to be this big!

- **Impact on Planet Earth's river systems**: Planet Earth's river systems are also failing to support the demands placed on them for drinking water, agriculture and industry. For example:

---

11   As reported on the UNEP website: www.na.unep.net.

- **Egypt**: The River Nile now only just makes it to the sea, largely due to the impact of the Aswan Dam. Before the dam was built, 32 billion cubic metres of water reached the Mediterranean Sea each year. Now due to irrigation, evaporation and other demands, less than two billion cubic meters arrive.[12]
- **Pakistan**: The Indus River, the country's main water supply, is starting to run dry in the lower reaches. At the same time, Pakistan's population is forecast to increase to over 300 million by 2050, creating further demands on already diminishing supplies.
- **South East Asia**: The supply of water to Mekong River is being impacted by dams built in other countries, such as China, which are in turn affecting downstream countries such as Cambodia, Laos, Thailand and Vietnam. With a combined population of over 150 million people, limited water suppliers are damaging relationships and increasing tensions between these countries.
- **USA**: The Colorado River which supplies the US states of Colorado, Utah, Arizona, Nevada and California now runs out before it reaches the sea.

- **Melting glaciers**: Increasing temperature on Planet Earth due to Climate Change is impacting the planet's glaciers, which are shrinking or entirely disappearing at an increasingly alarming rate. As the planet's glacier system is also non-renewable, once the world's glaciers have melted, availability of regular water supplies to many major population regions such as the Ganges basin will greatly diminish.
- **The increasing demands of cities**: Over half the world's population now lives in cities. There is therefore an increasing demand from the cities to supply the expanding urban populations. As cities are economically much stronger than farmers, they can therefore buy up the water rights much more easily. However, if farmers can then no longer supply the food that the cities need, no one wins – the farming communities lose their livelihood and the cities that gain the water lose a food supply. A number of examples highlight this issue from Asia to the Americas:
    - **South Korea**: A World Bank study has shown that the growth in residential and industrial water use could reduce the supply available for agriculture in South Korea by almost half in the next 10 to 15 years, from 13 billion to 7 billion tons, significantly impacting food production capabilities.
    - **USA**: The cities of San Diego, Los Angeles, Las Vegas, Denver and El Paso are all buying up the water rights of the surrounding countryside, Farmers no longer have enough water for irrigation to maintain existing food production levels.[13]
- **Increasing cross borders pressures on water supplies**: The issues are not only restricted to cities and farming communities. As major river systems flow through many countries, as we have already seen, the actions of one country can have dramatic effects on the countries downstream, as seen with the Nile and the Mekong.
- **Potential damage to industry**: Semi-conductor companies use vast quantities of clean water. Eleven of the world's largest semiconductor factories are located in the Asia-Pacific region where the risks of water shortages are high.
- **The impacts of bottled water**: Plastic water bottles not only consume significant energy but are a major source of pollution. The scale is alarming: in 2004, 28 billion bottles were produced. It is estimated that 1,500 plastic water bottles end up in the

---

12 As reported in *Plan B 3.0*, Brown, L. p. 76, Earth Policy Institute, W. W. Norton & Company. 500 Fifth Avenue, N.Y. 10110.
13 As reported in *Plan B 3.0*, Brown, L. p. 51, Earth Policy Institute, W. W. Norton & Company. 500 Fifth Avenue, N.Y. 10110.

garbage or the ocean every second. Each one of us with clean tap water available should stop using bottled water. Large-scale waste areas are now being discovered in the earth's ocean systems, with plastic deposits at times up to ten feet deep and twice the size of France in geographical coverage.

- **Climate Change**: The changing weather patterns driven by increasing Climate Change will make water management even more difficult with greater swings between drought and floods. Further, Climate Change mitigation programs such as the production of bio-fuels, often driven by governments subsidies, will further increase demands on often diminishing water supplies.

MEXICO CITY: A CASE STUDY OF INTERCONNECTIVITY

Mexico City is an example of the interconnectivity of the challenges and how they interact to impact on the supply of water to the city. With the city's population increasing over sixfold in the second half of the twentieth century, the demands on the water supplies have been stretched to their limits. Water consumption from aquifers is greater than their replenishment rate – meaning there is no long-term sustainability of supplies.

Further, due to the expanding urbanisation, the new roads and buildings have created a concrete barrier so that rain is no longer able to flow from the earth to the aquifers but is diverted to sewerage drains and wasted. Poor constructions also results in many water pipes leaking with as much as 40 per cent of drinking water lost.

The current position is desperate and getting worse, with aquifers reaching dangerously low levels. Water rationing has been introduced and inevitably water shortage will in turn create a new range of health problems.

Solutions and Best Practice: Water Supplies

Figure 34 **Yes, we've got plenty of reserves!**

OPPORTUNITIES

Efficiencies

There is a significant opportunity presented by the inefficiencies in water management. Current water usage in many areas from household, commercial and agricultural usage is often very inefficient. The EU estimates that over 40 per cent of water used in the EU is wasted.[14] However, it is worth looking first at where water is being used, and based on this, where strategies could be focused to increase efficiencies. From this chart, it is clear that the key priorities are agricultural, industrial and commercial usage, where an estimated 90 per cent of the human water supply is being used.

BEST PRACTICES

- **Advanced technologies**: Technologies to extract fresh water from salt water are improving and have an important future contribution. However, they require high levels of energy, potentially increasing CO_2 emissions.[15] Again, we are faced with the challenge of interconnectivity.
- **Integrated water management**: Better integrated water management is essential to effectively manage diminishing water supplies. The Pacific Institute states that by merely using current water saving practices, California, a water-poor state, could meet its needs for decades to come.
- **Water harvesting**: Collection of water that would otherwise evaporate or run off could significantly increase water supplies. Redirection of rainwater from sewers to water collection could make a significant difference in cities such as Mexico City, which are already facing major water shortages.
- **Improved agricultural practices**: The use of treated waste water and more efficient ways of using water for crop production provide important opportunities for water conservation. The International Commission on Irrigation and Drainage say that best practice irrigation systems can improve water efficiency by 30 per cent.
- **Pricing**: Historically water has rarely been priced in ways that reflect supply and demand. Indeed, it is often treated as a free resource to which all have a right. Creative pricing mechanisms can provide incentives for the more efficient use of water but have proved difficult to design. It is not always easy to define secure legal title rights to the water and difficult to accurately measure traded volumes. Further, wealthier city populations can often outbid farmers for water. The ten year drought in Australia has forced changes. Australian farmers have the right to a certain amount of free water. If it is more than they need they can sell the rights. This has proved effective and trading in this manner has doubled water productivity through a range of practices including farmers switching to less water intensive crops.
- **Research**: In March 2009, General Electric invested $1 million to establish a new research facility in association with the National University of Singapore. The resulting Singapore Water Technology Centre will develop new solutions for low-energy sea

---

14 As reported on www.euractiv.com/en/sustainability/water-business-sustainability/article-173757.
15 As reported in *The Economist*, p. 6, 5 June, 2008.

water desalination, water reclamation and improved approaches and technologies for water efficiency.

The Role of Government, Business and People

GOVERNMENT

Government has a key role to play in regulating water usage to achieve optimum levels of sustainability. This can be achieved through the setting of standards and regulations regarding water usage as well as controlling the cost of obtaining water.

- **International cooperative agreements**: Cooperative efforts between governments to establish sustainable water supplies are essential if conflicts over water supplies are to be avoided. For example, Ethiopia, Sudan and Egypt have already put an agreement in place which provides a best practice model for other countries and regions. These countries are working together to develop cooperative water management approaches to achieve more efficient and equitable use of their common major water source – The Nile. With the population forecast to increase significantly in all of these countries this cooperative approach is key to long-term sustainability of supplies.

    In spite of their political differences, India and China are now monitoring the Himalayan glaciers for melting due to Climate Change. These glaciers feed seven of the worlds greatest rivers, including the Ganges and the Yangtze. They supply water to about 40 per cent of the world's population.

    The EU has also undertaken a number of 'best practice' initiatives to encourage cooperation between countries and the critical strategies for the effective management of water.
    - **Water use in cities**: With over 80 per cent of people in the EU living in cities, the critical issue of supplying water in metropolitan areas is one where cooperative strategies are being developed.[16]
    - **Ground water and river resources**: Over 15 per cent of Europe's total water needs come from these sources, so initiatives are being developed for the sustainable management of these groundwater supplies through cooperation between various countries.
    - **Water purification systems**: Cooperation between countries including France, Spain, Egypt, Morocco, Israel and Austria is taking place to develop plans focused on drinking water and waste water purification.
    - **Water and ecosystems in regional development**: Plans are being developed for the Okavango River and Delta in Africa by a joint group of researchers from the UK, Netherlands, Namibia, Sweden and South Africa to find a balance between societal needs on the one hand and environmental needs on the other.
    - **Efficiency standards and labelling for appliances**: The high degree of waste water can also be managed through the increase in efficiency standards for water usage and appliances.

---

16   As reported on www.emwis.net/documents/meetings/events/selected-events-5th-world-water-forum-istanbul-36778.

BUSINESS

With an estimated 90 per cent of the global water supply being consumed in either food production or commercial usage, business has a critical role to play in developing and implementing strategies to improve sustainable water usage. In both of these areas, there is already best practice that can be expanded to a more global scale.

- **Food production and agriculture**: Best practice water efficiency usage is found in countries such as Israel, Jordan and Cyprus – countries with historically low water supplies which achieve water efficiency levels of between 50 and 60 per cent. This is far higher than countries such as India, Pakistan, Mexico and the Philippines where water efficiency is only between 25–50 per cent.[17] This best practice is being achieved by changing from traditional furrow irrigation, which is highly inefficient due to evaporation loss and loss to the soil, to drip irrigation, which is one of the most water efficient methods, increasing water efficiency levels by over 50 per cent.
- **Industrial and commercial usage**: Companies which have traditionally used water-based disposal systems, such as paper and pulp, industrial laundries and metal finishing are introducing best practice 'closed loop systems' for disposal and dispersal of industrial waste.[18]

PEOPLE

The third area of focus is household usage, where approximately 10 per cent of the world's fresh water supplies are used. Individuals can make an important difference in their household efficiency through application of simple but effective best practices.

Household usage: With over 50 per cent of household consumption being for toilets, showers and washing – there are a number of highly effective best practices which can be readily implemented.

- **Shower heads**: Changing from a showerhead flowing at 18 litres per minute to one flowing at 9 litres.
- **Washing machines**: Horizontal axis or front-loading washing machines use 40 per cent less water than traditional top-loading machines, as well as requiring less energy. One of the benefits of interconnectivity.
- **Toilets**: A traditional toilet uses over 22 litres of water to flush whilst the maximum for new toilet in the US is 6 litres. However, best practice dual flush technology, developed in Australia, where water shortages are high on the agenda, can reduce this usage by a further 20-30 per cent.[19]

Simply replacing household appliances with more efficient ones over the next 10 to 20 years, or sooner if feasible, could reduce household usage by 40-50 per cent.

---

17   As reported in *Plan B 3.0*, Brown, L. p. 180, Earth Policy Institute, W W Norton & Company. 500 Fifth Avenue, N.Y. 10110.
18   As reported in the OEDC website: www.oecd.org/dataoecd/45/27/1895218.pdf.
19   As reported in The World's Water 2004–2005: The biannual report on fresh water resources, Gleick, P., Washington DC, Island Press, 2004.

The Current Situation – Food Supplies

- **Higher population requires more food**: Despite these diminishing water supplies it has been estimated that farmers will need to grow 50 per cent more crops by 2030 in order to meet the requirements of forecast population growth and to avoid potential food shortage.[20]

    This challenge is not new, and was famously highlighted in with the publication of Paul Ehrich's *The Population Bomb* in 1968. Over the following decades this 'population bomb' was temporarily defused through the spread of the Green Revolution techniques pioneered by Norman Borlag, and introduced with dramatic results into key population growth regions such as Mexico, Pakistan and India (see box insert).

    The spread of these techniques continues with significant current focus in Africa. One of the key challenges, however, in the continued expansion of these techniques is the consumption of water supplies, which as we have already seen, in many areas in Planet Earth are non-sustainable. This is already having an impact in previously highly successful Green Revolution areas such as India. Increased food production will therefore need to be combined with increased water efficiency and supply capabilities, as well as addressing the challenges of unsustainable population growth, if they are to be successful.

- **Impact of diminishing water supplies on food production**: It is clear, therefore, that the supply of food is dangerously dependent on the previously discussed diminishing supplies of water to support the required increase in agriculture. For example:
    - **USA**: The Ogallala aquifer, the largest underground water source in the USA, and a prime source of water for food production in the Great Plains Region, has dropped more than 100 feet since pumping first started. Water is being withdrawn at an unsustainable level of 10 to 40 times faster than it is being replenished.[21]
    - **Other countries**: Rapidly depleting underground water supplies could also reduce food production for China, India, Pakistan, Mexico and nearly all the countries of the Middle East and North Africa. The challenge for those countries which are depleting these supplies is to reduce the use of water and thus stabilise water tables.

- **Lifestyle changes**: As incomes rise there is a shift from vegetarian diets to meat-based diets. A key example of this is the consumption of meat in China, which has more than doubled from 20 kg in 1985 to the current rate of 50 kg per person. A kilogram of meat, however, requires 15,000 litres of water, compared to wheat which only requires 1,000 litres of water. These types of lifestyle changes are therefore further increasing the demand on the world's diminishing water supplies.

- **Crop production reduction due to global warming**: Climate Change potentially reduces the ability of the human race to produce the three main crops which provide the backbone of food production – wheat, rice and corn. For example, a 1 degree centigrade rise has been shown to reduce crop yield from between 5 and 7 per cent in major crops, such as rice.[22] A forecast increase of 2 degrees in global temperature, as a result of global warming, could significantly reduce the ability of the crops to be pollinated to produce the grains that are harvested for food.

---

20 As reported in *Time*, UN Secretary General Ban Ki-Moon, p. 44, 16 June, 2008.
21 As reported on the HRW website: http://go.hrw.com/resources/go_sc/bpe.
22 As reported on http://www.icrisat.org/Journal/SpecialProject/sp8.pdf.

Water and Food Supplies   79

Figure 35   Not tonight Josephine, it's too hot

- **Impact of droughts**: The negative effects of Climate Change on food production are further illustrated by the 2007 droughts in Australia. It has been estimated that production of wheat and rice crops was reduced by over 40 per cent from the normal production levels due to the prolonged drought which has been linked to Climate Change.[23] The UN World Food Programme has identified the 'silent tsunami' of drought which could potentially impact up to 100 million people around the world.[24]
- **Impact of flooding**: Climate Change has already caused an increase in the frequency and severity of storms and flooding in many regions, further disrupting and damaging food production.
- **Impact of rising sea levels**: Due to global warming, and the influence this is having on the melting of ice at the poles, in Greenland and in glaciers, sea levels are already forecast to rise by up to a metre. In Bangladesh, for instance, this rise in the ocean levels would cover over half of the rice growing land. With a population currently of 142 million, forecast to grow to over 230 million by 2050, the significance of these sea level changes is dramatic and demonstrates the close interconnectivity between Climate Change, the demands of increasing population and food production capabilities.
- **Fisheries**: Fisheries, a key source of food from rivers, lakes and the ocean, are also under significant threat. Since the 1950s, human seafood consumption has increased from 19 million tons in 1950 to 93 million tons in 1997.[25] This diagram from the

---

23   As reported in *The Guardian*, 3 November, 2007.
24   As reported on *The Sun Herald* website: www.news.com.au/heraldsun/story/0,21985,23589032-2862,00.html.
25   As reported on the One World website: http://us.oneworld.net/article/358611-finding-nemo-soon-impossible.

UN Environmental Programme highlights the cumulative impacts on the marine environment,[26] from over-harvesting, damaging the ocean beds and habitat loss due to coastal development.

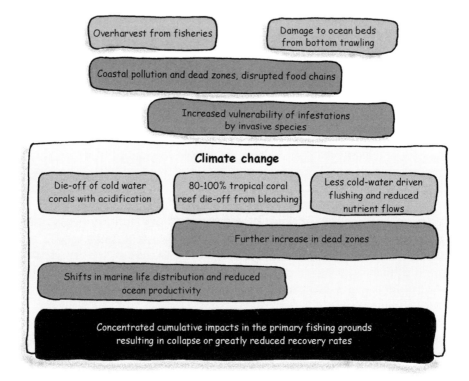

Figure 36   Cumulative damage to the marine environment

- **Sustainable capacity**: According to the Food and Agricultural Organisation (FAO) for the UN, consumption of fish and fishery products increased by 260 per cent between 1969 and 2003. In 1950, just 15 per cent of stocks were over fished. By 2003, it was estimated that 70 per cent of stocks were over fished or had totally collapsed,[27] with the Atlantic Ocean production peaking in the early 1970s, the Pacific Ocean in 1980s and the Indian Ocean in the 1990s.
- **Africa**: African countries are selling fishing licences which allow EU fishing boats to fish off the coast of countries such as Senegal, Mauritius, Morocco, Guinea, Bissau and Cape Verde. There is concern that this is not only impacting food production in these regions, but also destroying the livelihood and traditional food supplies of local fishermen.
- **European Union**: The impact in the EU is dramatic, with 88 per cent of the EU's stocks already over fished. Ninety-three per cent of North Sea cod are caught before they can breed, clearly impacting the sustainability of these fisheries.

---

26   Used with permission: Hugo Ahlenius, UNEP/GRID-Arundel.
27   As reported on *The New York Times* website: http://www.nytimes.com/2008/11/16/weekinreview/16bittman.html.

- **Canada**: Lack of sustainable approaches to fisheries lead to the 500-year-old cod fishery failing in the early 1990s – putting some 40,000 fish catching and production workers out of work.
- **Fish farming reduces crop supplies**: As growing demand for seafood can no longer be satisfied by the expanding ocean fish catch, there has been an expansion in fish farming. Seafood being produced through aquaculture has increased from 12 per cent of world fishery production in 1984, to 38 per cent by 2003. FAO expect aquiculture will exceed 50 per cent of the world's edible seafood supplies by 2025. But once fish are put in ponds or cages they have to be fed which puts further pressure on grain production.
- **Food for all**: It is estimated by the UN FAO that over 850 million of the people on Planet Earth are chronically hungry. The majority of these people are in Sub-Saharan Africa but many developing countries are also being affected. The impact is that these people are not getting enough food to achieve full physical and mental development and maintain appropriate levels of physical activity. This malnutrition damages the young who are most vulnerable during their rapid physical and mental development. For example:
    - **Ethiopia**: Almost 50 per cent of children are undernourished, with population estimates expected to double in the next 40 years.
    - **India and Bangladesh**: Almost half of the children under five are underweight and malnourished.
    - **Nigeria**: In Nigeria, the figure is 31 per cent, with population estimated to also more than double in the next 40 years.
- **The water/food production 'bubble'**: As the water available to supply agriculture in many parts of the world decreases, the ability of these regions to produce their own food reduces. These regions are therefore forced to purchase more grain. As this is produced in other regions this in turn increases the use of their available water supplies to fill the gap – and also increases the carbon footprint from transportation between regions. Again, we see the close interconnectivity of the challenges.

Figure 37   The food/water bubble

- **Energy – food interconnectivity**: This interconnectivity is further demonstrated by the production of bio-fuels, which is the largest source of new demand in decades.

This is having an impact in driving up agricultural commodity prices,[28] which in turn reduces the ability to source and supply food for the chronically hungry.

Solutions and Best Practice – Food Supplies: The Role of Government, Business and People

GOVERNMENT

The cross impact of the Global Challenges on food supplies means that effective cooperation and coordination between international and national governments will be key moving forward in ensuring adequate and sustainable food supplies for the global population.

There are, however, increasing examples of where this type of cooperation is taking place with very positive impacts. Nowhere is the principle of Thinking Globally and Acting Locally more important. A good example of this is the global coordination being translated into local action in addressing the requirement of long-term sustainable fisheries.

- **Regenerating fisheries**: The world's fisheries have been significantly depleted over the last 50 years, yet they remain a vital source of food both locally and globally. The establishment of a global network of marine reserves (essentially nurseries) covering the key breeding grounds (30 per cent of the world's oceans) could significantly increase the availability of food resources from oceans, lakes and rivers and will make a major contribution. This has commenced and is already making a significant impact on replenishing fisheries in these regions.
- **International cooperation on ocean dead zones**: Another problem area is the number of dead zones which have been created in the world's oceans due to the flow of sewage and excess fertiliser. A UN report indicated that the number of low-oxygen dead zones has increased from 149 to more than 200 in the last two years and often these are located closest to the populations that would benefit the most from fisheries. The worst example is the 'Great Pacific Garbage Patch' where an area twice the size of France and 10 metres in depth has floating plastic debris as its main ingredient. Best practices are being developed to achieve the reversal of this increase in dead zones in the world's oceans and this needs to be expanded as a matter of urgency.[29]
- **Investment in developing a nation's food production**: Investment is increasing. For example, IFC, a member of the World Bank Group, is providing $75 million, its largest investment, in agribusiness to support farmers and food production in emerging markets to increase global food supply.
- **A new initiative on agricultural methods**: In July 2009, the Worldwatch Institute launched a major initiative to assess the agricultural methods' impacts on sustainability and productivity. Funded by the Bill and Melinda Gates Foundation with $1.3 million, the project will focus on Sub-Saharan Africa. Amongst practical solutions being examined are:

---

28 As reported on OECD website: www.oecd.org/dataoecd/1/36/41227216.pdf.
29 As reported on the Botany UWC website: www.botany.uwc.ac.za/PSSA/articles/features/no63.htm.

- Adding nitrogen-fixing plants into crop rotations as a low-cost solution for enriching soil and breaking weed and pest cycles.
- Overcoming fresh water shortages with rain harvesting, efficient irrigation, micro dams and cover cropping.
- Strengthening local breeding capacity, including the use of farmer-run seed banks and genetic markers of important crop traits.
- Tapping international carbon credit markets to reward farmers for enriching their solids and planting carbon carbon-sequestering tree crops.
- Involving women farmers in decision making at all levels.

This initiative is endorsed by many African governments.[30]

A NEW GREEN AGRICULTURAL REVOLUTION

The term Green Revolution refers to the transformation of agricultural productivity which started in Mexico in 1943 when new varieties of wheat were developed and supported by fertilisers to feed a rapidly growing population. Funding came from the Rockefeller Foundation and Ford Foundation and the scientific input through a new agricultural research station. In 1943 Mexico imported half its wheat, but by 1956 the country was self-sufficient and by 1964 it exported half a million tons of wheat.

Building on that experience, and led by Nobel Prize winner Norman Borlaug, the Green Revolution spread to Asia, the Indian sub-continent and later to Latin America. These were territories where in the 1960s there were legitimate fears of long-term famine, exacerbated by population growth. A prospect with major human, social and economic consequences. The improvements were dramatic. Not only was famine substantially averted, but a prolonged period of low-cost and plentiful food supplies followed.

For example, India, faced in 1961 with a serious famine, adopted a new rice variety which, when coupled with better irrigation and fertilisers, increased rice yields from two tons per hectare to six tons per hectare by the mid 1990s In Indonesia, annual rice production increased from 3.7 tons to 7.7 tons in 20 years. Later, new high-yielding varieties included rice and wheat, sorghum, cassava and beans

This Green Revolution had four key elements. Scientists developed new seeds which were not just high yielding but also resistant to insects and drought. Water, so often wasted, was better managed through new dams and reservoirs and irrigation. Pesticides and fertiliser were used extensively for the first time. Agricultural research stations provided local knowledge, education and training and trial work.

The dramatic success of this work led to complacency. Aid funds dedicated to agricultural developments were reduced and the focus turned more to cost reduction instead of increasing yields. This resulted in shortages persisting, which were then addressed in the old fashioned way. For example, in 2003 the US gave Ethiopia more than $500 million in American-grown grain but only $5 million in agricultural development aid so farmers could help themselves.

---

30 As reported in Worldwatch Institute Press Release, July 8, 2009, Washington D.C.

The focus turned more to cost reduction rather than high yields. Global grain reserves dropped in 2007/8 to their lowest level for 30 years.

Agricultural productivity improvements slowed down but population continued to grow. The higher costs of fertiliser and seed led to high food prices and, inevitably, the poorest people were hardest hit. Food riots broke out. This concentrated the mind of the international community on the importance of food productivity.

However, thoughts of just repeating the first Green Revolution are misconceived. Energy-intensive farming is costly as gas and oil prices increase. Whilst some see GM crops as the new hope for the future there is an influential group which points out the risks. High-yield crops are water intensive yet water is becoming a seriously scarce resource. Over use of pesticides and fertiliser has led to seepage into rivers and lakes causing serious environmental and health problems.

The thrust of the new Green Revolution is for more organic farming, less use of fertiliser, development of risk-free seeds, improved agricultural processes and practices, modern drip irrigation; non-till methods and micro-credit facilities for small farmers. A concern for sustainable improvements in agricultural productivity.

The Challenge of Africa

Africa faces the most serious food problems, not least because the first Green Revolution which was so successful in Asia and elsewhere passed Africa by. Over 90 per cent of Sub-Saharan Africa is rain fed and not irrigated and is therefore vulnerable to Climate Change. African countries are too poor to buy sufficient fertilisers which have doubled in price in the last few years. The UN Food and Agriculture Organisation reports that Africa has to import about 28 per cent of its calorific requirements – wheat, oils and rice – all of which have been increasing in price on the international market. Moreover, because the basic infrastructures are inadequate, transporting imported food and fertiliser inland from the ports greatly increases costs to the farmer.

Two other issues increase the urgency of the challenge. The population of Sub-Saharan Africa will, according to the UN estimate, double by 2050. At the same time HIV/AIDS is decimating the most productive adult members of society between the ages of 18 and 45.

Solving these problems is complex since it involves investment in infrastructure, aid focused on agricultural development not the provision of food itself, no till farming, the establishment of many more agricultural research stations, organic farming, and better irrigation. To say nothing of the need for non-corrupt effective governance.

There is hope. Twelve African countries, led by the Alliance for a Green Revolution in Africa in partnership with others, have produced practical plans to double Africa's rice production by 2018. Kofi Annan, then Head of the UN, put it bluntly to Africa's heads of state and international development specialists in 2004:

'Nearly a third of the men and women and children in Sub-Saharan Africa are severely undernourished. Africa is the only continent where child malnutrition is getting worse rather

> than better. We are here together to discuss one of the most serious problems on earth: the plague of hunger that has blighted hundreds of millions of African lives – and will continue to do so unless we act with greater purpose and urgency.'

Malawi, rejecting the policy requirement of World Bank subsidies in 2004, financed 1.4 million small farmers with a $55 million fertiliser and seed programme. Corn production was doubled. The message to aid donors was loud and clear: Don't send us your subsided surplus food. Send us seed and fertilisers and agricultural experts and we will do the rest.

Conclusion

The Food and Agricultural Organisation in 2009 projected a rise of 915 million hungry people in 2008 to 1.02 billion in 2009, mainly in developing countries. That is about of one-sixth of humanity undernourished.

We have yet to respond to the challenge set out in the Nobel lecture given by Norman Borlaug in 2007:

> 'Man can and must prevent the tragedy of famine in the future instead of merely trying with pious regret to salvage the human wreckage of the famine, as he has so often done in the past. We will be guilty of criminal negligence, without extenuation, if we permit future famines.'

BUSINESS

Business is playing a key role in increasing production to meet the earth's requirements, and there are many best practices that can be expanded and rolled out globally. For example, business can strengthen the food value chain as shown in the World Economic Forum, Business role in strengthening Food Value Chains Report, p. 12 (www.weforum.org).

Table 5   **The business role in strengthening food value chains**[a]

	Agricultural Production	**Storage and Transport**	**Selling and Trading**	**Processing and Packaging**	**Retail Distribution**
Challenges	Farmers lack access to inputs; they need local distributors, financial services and training for improved seeds, fertiliser, and water management.	Storage facilities are inaccessible or poorly managed. Transport costs are very high and infrastructure is very poor.	Prices fluctuate and farmers lack access to price information and finance. High transaction costs and few market linkages for small producers. Irregular supply and quantity for large buyers.	Lack of facilities or capital for on-farm processing. Lack of commercial-scale processing facilities. Few appropriate packaging technologies. Lack of food fortification regulation and incentives.	Lack of retail outlets for essential products and services in poor regions. Lack of business linkages to help existing retailers expand or diversify.

Table 5   **The business role in strengthening food value chains**[a] *concluded*

	Agricultural Production	Storage and Transport	Selling and Trading	Processing and Packaging	Retail Distribution
Effective Business Models	Agro-dealers expansion (through training and finance). Consumer financial services (Vouchers or microfinance). Irrigation equipment and services.	Provision of storage services and infrastructure.	Sourcing from small farmers, producer groups and cereal banks. Market information services via telecom/IT. Financial services for entrepreneurs.	Small-scale processing for local markets. Commercial-scale processing for national and regional markets. Appropriate packaging. Food fortification.	Retailer expansion (through training, equipping and finance). Retailer diversification (expanding products and services offered). Transport/delivery service.
Improving capacity and the enabling environment	Training, organising and financing producers to use input technologies and meet quality standards.	Improving infrastructure (storage, transport, energy and communications).	Strengthening entrepreneurs by providing business tools, services, investment and linkages.	Investing in locally-produced packaging and processing for competitive products.	Training, equipping and financing retailers to effectively market essential goods in poor communities.
	Strengthening regional growth by identifying and investing in commercially viable, regionally competitive products. Partnering with government and public institutions to set joint priorities for investing in growth, and to improve execution.				
Relevant Industries	Fertiliser, seed and equipment manufacturers. Banks. Retailers and agro-dealers.	Transportation services. Warehousing and logistics companies. Construction companies.	IT and telecoms. Banks and insurance. Wholesalers. Food manufacturers, processors and retailers.	Food and beverage manufacturers. Packaging firms. Bio-fuel processors. Millers.	Retailers. Transportation services and infrastructure. Banks. IT and telecom firms.

[a]   Reproduced with permission by World Economic Forum, Business role in strengthening Food Value Chains, p. 12.

- **Multicropping**: Multicropping provides an important best practice which can be used to increase the amount of food which can be produced in an area. As this practice is still only limited in use, there is a significant opportunity for government and business to work together to introduce and expand this best practice – and thus increase food production.
- **Nutritious food**: Businesses are making an important contribution through the development of highly nutritional products for poorer regions. For example, Grameen Danone Foods in Bangladesh have an affordable yogurt which meets the unique nutritional needs of Bangladeshi children. By the end of 2009 it is expected the yogurt will reach 300,000 children.

PEOPLE

Finally, there are many best practices which people can utilise, both in developing countries through the use of best practice farming techniques, and in developed counties through the selection of the food that is consumed.

- **Organic farming**: A major area of progress has been in the development and application of organic farming and multicrop techniques in Africa, which can be applied to other developing countries. Studies in Ethiopia have demonstrated that the use of modern organic farming techniques increases average crop yields across both drought and non-drought periods above those achieved by the use of fertilisers.[31] Further, the use of natural fertilisers not only provides protection from drought times but it also increases water conservation. This is a good example of the positive impact of interconnectivity. Experimental work in Nepal is demonstrating encouraging results.
- **Food selection**: In developed countries people can play an active part in the selection of food that is purchased. Purchase of appropriately certified and labelled food products, such as organic foods, contribute to the growth of these sustainable food supply businesses and drive innovation and change towards increased food sustainability. Similarly, understanding the carbon footprint of certain foods can also be important influencers of consumer choice. These positive consumer decisions can also make an obvious difference in areas such as Fair Trade, biodiversity and Climate Change.

---

31 As reported on the Twnside website: http://www.twnside.org.sg/title/jb19.doc.

CHAPTER 6

Executive Brief No. 5: Planet Sustainability and Biodiversity

'We are relentlessly taking over the planet, laying it to waste and eliminating most of our fellow species. Moreover, we're doing it much faster than the mass extinctions that came before. Every year, up to 30,000 species disappear due to human activity alone ... and we could lose half the Earth's species in this century. And, unlike previous extinctions, there is no hope that biodiversity will recover – since the cause of this devastation – us – is here to stay.'

Jerry Coyne and Hopi E. Hoekstra

Figure 38   The last tree?

SUMMARY

The rapid increase in the human population on Planet Earth is not only putting pressure on water and food resources but also on land, forests and animal species. Loss and degradation of these resources causes the loss of biodiversity.

However, we are becoming increasingly aware of the interconnectedness of all species – the 'web of life' – which means that the long-term implications of these losses are unpredictable.

90 Developing a Plan for the Planet

> They could have a profound impact on the sustainability of all life on Planet Earth – yet large-scale destruction continues.
>
> For these reasons, establishing sustainability objectives and strategies for all of these resources is essential in establishing the long-term viability of the human race.
>
> Much best practice is already emerging in the areas of forest protection, reforestation, reclaiming of deserts and protection of the diversity of species. The challenge is to globalise these best practices to ensure they are implemented in a coordinated and consistent manner – whilst working to stop further destruction of the existing, finite resources.

The Current Situation

- **A thin layer supporting civilisation**: A thin layer of topsoil averaging about three feet has provided the foundation of human civilisation. On this thin layer of soil has been built the agrarian revolution from which modern civilisation has developed. However, this thin layer of soil is eroding. The world's cropland is losing topsoil through soil erosion faster than new soil is forming, thus reducing Planet Earth's land productivity[1] and therefore the ability of this thin layer to continue to support the growing human population.
- **Loss of other species**: Not only is the soil eroding, we are also rapidly losing other species. In the second half of the twentieth century, Planet Earth lost over 300,000 species, with present species disappearing between 100 and 1000 times faster than before Homo Sapiens evolved. The proportion of birds, mammals and fish that are vulnerable or in immediate danger of extinction has now reached double digits. For example:
    - 12 per cent of the world's 10,000 bird species are endangered.
    - 23 per cent of the world's 4,776 mammal species are endangered.
    - 46 per cent of the world's fish species are endangered.

Figure 39   Hey, did you know that the bees used to do all of this for nothing?

---
1   As reported in the *Financial Times*, 17 July, 2007.

- **Cross impacts of loss of biodiversity**: As this biodiversity of plant and animal species is threatened, key elements of the earth's life-support system, such as soil formation, are further damaged. A vicious circle of environmental destruction. A good example is the impact we are already seeing on the bee population – which in turn affects the pollination of valuable crops as well as other plant species (see box insert). The biological control of pests and diseases is also another area impacted by the loss of biodiversity. Edward O. Wilson, a famous entomologist says, 'Destroying a tropical rain forest and other species of rich ecosystems for profit is like burning all the paintings in the Louvre to cook dinner.'
- **Plants are pharmaceutical gold mines**: The bark of trees, for example, has given us quinine (the first cure for Malaria), taxol (a drug highly effective against ovarian and breast cancer) and aspirin. More than a quarter of the medicines on our pharmacy shelves were originally derived from plants. The sap of the Madagascar periwinkle contains more than 70 useful alkaloids, including vincristine, a powerful anti-cancer drug. Of the roughly 250,000 plant species on Planet Earth, fewer than 5 per cent have been screened for pharmaceutical properties. Who knows what life saving drugs remain to be discovered? Given current extinction rates, it's estimated that we are losing one valuable drug every two years.[2]
    - **Loss of forests**: It is estimated that almost as much man-made deforestation has occurred since 1950 as occurred in the entire period of human civilisation previous to this.[3] Over the last 100 years, the earth's forested area has been reduced by more than 20 per cent to under 4 billion hectares. The demand for wood for fuel and timber products, combined with advances in technology, has led to the more rapid harvesting of forest products. This is resulting in major areas of deforestation in many of the last remaining virgin forest regions. In Malawi, a country of 13 million people in East Africa, forest resources had been reduced from 47 per cent of the total land area to around 28 per cent by 2000.[4]
    - **Large-scale deforestation**: Large-scale deforestation is taking place on a global scale, covering Africa, Asia and South America, home to over 75 per cent of the world's 6.8 billion people – and it is having a dramatic impact on the forests themselves – as well as the plant and animal species that inhabit these regions.

        In New Guinea, Indonesia and Thailand, forest areas are being felled to support the development of the palm oil bio-fuel industry as well as to supply wood products. In Thailand in 2004, devastating flooding and landslides, caused by rampant deforestation, killed hundreds of people. Similar devastating flooding has occurred in China as a result of deforestation of the river regions.
    - **The Amazon rainforest**: 20 per cent of the earth's oxygen is produced by the Amazon rainforest, yet it is being systematically destroyed. It is estimated that almost all of the Atlantic rainforest in Brazil has been lost. The Amazon rainforest itself, roughly the size of Europe and largely intact prior to 1970, has

---

2   As reported in *The New Republic*, Coyne, J. and Hoekstra, M., 24 September, 2007.

3   As reported in *Deforesting the Earth*, Williams, M., Chicago University Press, Chicago, p. xvi.

4   The Ministry of Natural Resources and Environmental Affairs said in its 2002 State of the Environment Report, Unchecked Deforestation Endangers Malawi Ecosystems By Charles Mkoka, Environment News Services as reported in www.ens-newswire.com/ens/nov2004/2004-11-16-04.asp

Figure 40 The treasures of our rainforests

  been reduced by over 20 per cent in the last 40 years. This has been impacted by the increased demand for bio-fuels and the adverse damage to the Amazon forests from neighbouring countries such as Peru. For instance, the USA subsidies for corn ethanol, have contributed to forest clearing in the Brazilian Amazon.
  - **Siberia**: Large areas of Siberian forest are being devastated by advanced production techniques.
  - **Impact on the food cycle**: This deforestation is also contributing to the expansion of deserts and therefore further decreasing the land available for food production – a vicious cycle.
- **Counting the cost**: The economic cost is also significant, and not always fully understood or considered. A 2009 report funded by the European Commission, 'The Economics of Ecosystems and Biodiversity', states that the annual cost of forest loss is between $2 trillion and $5 trillion. The figure has been developed by adding the value of the various ecosystems the forest provides – such as cleaning water and carbon sequestration.[5] It is useful to compare these figures with the costs of addressing the 2007–2009 financial crisis.
- **Desertification**: Growing population, the increasing demand for resources and expanding demands of livestock is increasing desertification in many regions. This includes China, and other countries in the region such as Iran and Afghanistan, many African countries including Senegal, Sudan, Ethiopia, as well as Latin American countries such as Mexico and Brazil. For example:
  - **Rajistan Desert, India**: Up to 100 villages have been submerged by windblown dust and sand.[6]
  - **Mexico**: 65 per cent of the national land area of Mexico has been degraded, contributing to rural urban migration of 700,000–900,000 people each year from Mexico's drylands.[7]

---

5 As reported on the BMU website: http://www.bmu.de/files/pdfs/allgemein/application/pdf/teeb_phase2_hg_en.pdf.

6 As reported on Wikipedia website: http://en.wikipedia.org/wiki/Registan_Desert.

7 As reported on the Lada website: lada.virtualcentre.org/eims/download.asp, Impact of land degradation on ecological services besides agricultural production.

Figure 41 **The treasures of our rainforests are rapidly disappearing**

- **Wildlife**: This combination of the growing human population, expansion of crop and farmland and destruction of forests, lakes and rivers is having a dramatic impact on the other inhabitants of Planet Earth. Every year, up to 30,000 species disappear due to human activity alone. Following are a few examples:
    - **Caribbean**: It is estimated that the numbers of leatherback turtles have decreased from over 100,000 adult females in the early 1980s to 30,000 in 1996 – a reduction of up to 70 per cent.[8]
    - **China**: A search for the remaining Yellow River dolphins in an effort to relocate them to a safe location, failed to find any, resulting in the species now being identified as possibly extinct.[9]

Figure 42 **So long, and thanks for all the fish…**

---

8 As reported on the Red List website: http://www.iucnredlist.org/details/6494, the Red List, *Dermochelys coriacea*.

9 As reported the *National Geographic* website: http://news.nationalgeographic.com/news/2006/12/061214-dolphin-extinct_2.html.

- **North America, Europe and South Africa**: Over one-third of the species of fresh water fish are threatened or endangered by over-fishing arising from rapid growth in population, pollution and damage to the river ecosystems.

THE PROBLEM OF THE BEES

'If the bee disappears off the face of the globe then man would only have four years of life left.'
Albert Einstein

Sitting in the garden on a summer's day, listening to the drone of a honeybee as it moves from one flower to another, is a deep pleasure. It never occurs to us that bees are in crisis and indeed may become extinct. Still less do we think of bees in economic terms, crucial to the success of agriculture.

In fact, bees are dying at a concerning rate, mainly due to what is being termed Colony Collapse Disorder (CCD). This is a phenomenon in which worker bees disappear and the colony dies. So far it has primarily affected domestic commercial honey bees. Some deaths, in the range of 20 per cent, are normal and expected. However, current losses are significantly higher than this and range from 30–60 per cent on the West Coast of the USA, to 70 per cent in Texas. In Europe, CCD has spread to Germany, Switzerland, Spain, Portugal, Italy and Spain. In the UK, John Chapple, one of London's biggest beekeepers, reported that 23 of his 40 hives were abruptly abandoned.

The danger is very simple: no pollination, no crops. Most of the pollination of more than 90 commercial crops in the US is provided by honeybees – with the value to agriculture output estimated at £8 billion. 'Every third bite we consume in our diet in the US is dependent on the honeybee to pollinate that food,' said Zac Browning, Vice President of the American Beekeeping Federation. In Britain, ten crops alone, ranging from apples and pears to oilseed rape, are worth £165 million per annum.

The causes of CCD are uncertain and theories range from the presence of new viruses to the damage caused by heavy use of chemicals in intensive modern farming. As the problem is better recognised there has been some increase into research into the basic causes of CCD, however, the investment in such research is far too small, given the scale of the problem and its economic consequences. Lord Rooker in November 2007 informed the British House of Lords that if things went on as they were, the honeybee would be extinct in ten years. Many experts believe eight years is more realistic.

The response to this crisis indicates a sense of desperation. British bee keepers are importing captive-bred bees, many of them southern European species. The total is in the order of 10,000 nests, each containing a queen and 200 workers, at a cost of £50. However, as well as the economic impact, there are risks of bringing in alien species.

In Sichuan province, China, the most important crop is pears which depend on pollination. An overuse of pesticides has destroyed the bee population. Literally thousands of villagers have been pollinating by hand, using paint brushes. Not a sustainable model for the rest of the world!

Opportunites and Best Practice

'The one process that will take millions of years to correct is the loss of genetic and species diversity by the destruction of natural habitats. This is the folly that our descendents are least likely to forgive us for.'

Edward O. Wilson

OPPORTUNITIES

- **Carbon Trading**: Carbon Trading projects are beginning to have a positive influence on the saving of vital regions of forests.[10] Fourteen tropical and sub-tropical countries have been selected as the first members of a new partnership to combat tropical deforestation and Climate Change. Funding from the Forest Carbon Partnership facility will compensate the countries for reducing greenhouse gas emissions from deforestation and forest degradation.[11]
- **Marine parks**: Many countries are beginning to introduce best practice marine parks which are already encouraging the regeneration of fisheries in these areas. In 2009, South Africa announced its intention to establish the world's fifth largest marine reserve around the Prince Edward Islands in the Southern Ocean benefiting huge populations of seals, seabirds, killer whales, squid and many others.
- **Wildlife parks**: Some countries have found success with wildlife parks. For example, in some regions of India with their tiger management programmes.

Figure 43   We can rebuild the earth's forests!

---

10   As reported on the World Trade Review website: http://www.worldtradereview.com/news.asp, OECD report warns environmental action urgent, but affordable.

11   As reported on the Source Watch website: www.sourcewatch.org/index.php?title=Reducing_Emissions_from_Deforestation_and_Degradation.

- **Restoring the forests**: Protecting the earth's estimated four billion hectares of remaining forests and replanting those that have been lost is a key opportunity for restoring the earth's health.
- **The benefits of interconnectivity**: There is an opportunity presented by interconnectivity because many of the strategies that can be used to address climate, energy, water and food challenges can also contribute to restoring the long-term biodiversity of the planet.

The Role of Government, Business and People

GOVERNMENT

Clearly, the role of both international and national government is key in managing the issues of preserving and developing the planet's forests through regulation and forest management.

- **Reduce global deforestation**: Requiring both national and international government action, deforestation not only has significant impact on issues such a flooding and loss of biodiversity but it also reduces the earth's ability to reabsorb CO_2 from the atmosphere. Some excellent best practice has already been established. The Alliance for Forest Conservation and Sustainable Use has set targets to reduce the global deforestation rate to zero by 2020 with emphasis on proper certification for wood products. The benefit of certification is that it ties in well with offset programmes. Certified forests can become the beneficiary of appropriate offset funding but only certified forests.

    This provides important incentives for good forest management. Examples of where this is demonstrating clear benefits are the Guinean Rainforest which was saved through an offset programme.
- **Harvest from managed plantations**: There is an important opportunity for governments to both encourage and legislate for the replacement of the world timber harvest with wood from well managed plantations, rather than from natural forest land. The UN Food and Agriculture Organisation estimates that the harvest from managed plantations could more than double in the next three decades.
- **Refocus forest subsidies**: An important government best practice strategy is replacing subsidies for building logging roads with subsidies for the protection of forest and reforestation programmes on deforested and marginal land.
- **Reforestation**: Advances in modern agriculture and technology can enable large areas which have been previously lost to erosion and desertification to be reclaimed and turned into forest, benefiting the region through better flood control, biodiversity and agricultural land protection.

PROTECTING PLANET EARTH'S ANIMAL AND PLANT BIODIVERSITY

- **Water management**: It is essential that governments ensure that effective water management strategies are in place as they play a key role in protecting both fresh water and marine species.

- **Creation of reserves**: Significant progress is being made in this area. Over 10 per cent of the planet's land surface is now set aside in the form of protected reserves. Expanding the areas set aside for reserves, as well as increasing the protection in these areas, particularly in developing countries, needs to be a high priority and is reaping significant benefits where it is being successfully implemented.
- **Protecting endangered species**: Introduction of the Endangered Species Act in the US in 1973 is credited with already having saved a number of plant and animal species, including the bald eagle.[12] Expansion of this best practice on a global scale provides an important opportunity for protecting the biodiversity of the planet. This, however, will require coordination and cooperation on a global scale.

RECLAIMING PLANET EARTH'S DESERTS

- **Planting grass and trees on highly eroded land**: There are important examples of best practices that have already been developed in this area. The US provides an important example of the protection of areas with highly eroded soils from becoming wastelands. The 1930s dust bowls which developed following land degradation in the US stimulated the planting of tree shelter beds and planting of wheat on alternative strip beds to reduce wind erosion. These have proved to be highly effective and provide approaches which can be easily transferred to other countries and regions. Further, the introduction in the US of the Conservation Reserve Program in 1985, which included reduction of soil erosion through the overproduction of basic commodities, and the use of conservation practices and retirement of fragile cropland to grass, saw the reduction of soil erosion from 3.1 billion tons to 1.9 billion tons between 1982 and 1997.[13]
- **Conservation tillage**: The introduction of both non-till and minimum tillage has also proved effective in reducing the loss of topsoil. Instead of ploughing, discing or harrowing, farmers drill seeds through crop residues. 25 million hectares of non-till land are operating, particularly for the production of corn and soy beans. By 2006, 25 million hectares in Brazil, 20 million in Argentina, 13 million in Canada and 9 million in Australia will be using conservation tillage.[14] This also has the benefit of increasing water efficiency.
- **Building of green walls**: Concentration of perennial plantings such as orchards and vineyards, as well as tree planting in vulnerable regions, is another important best practice. There are a number of examples of where this is being successfully implemented.
    - **Africa**: The Green Wall Sahara Initiative, involving the planting of 300 million trees, will stretch from Senegal to Nigeria, aiming to halt the encroachment of the Sahara.[15]

---

12   US Fish and Wildlife Service, 'The Endangered Species Act of 1973' as reported on the FWS website: www.fws.gov/endangered, 'New Tool to fight global warming', Christian Science Monitor, 7 September, 2007.

13   US Fish and Wildlife Service, 'The Endangered Species Act of 1973' as reported on the FWS website: www.fws.gov/endangered, 'New Tool to fight global warming', Christian Science Monitor, 7 September, 2007.

14   US Fish and Wildlife Service, 'The Endangered Species Act of 1973' as reported on the FWS website: www.fws.gov/endangered, 'New Tool to fight global warming', Christian Science Monitor, 7 September, 2007.

15   As reported on the Sustainable Development website: http://www.sustdev.org/index.php?option=com_content&task=view&id=2028&Itemid=36.

- **Algeria**: The concentration of orchards and vineyards in southern regions to halt desertification of its cropland has been very effective.[16]
- **China**: The Green Wall is a project to plant 2,800 miles of trees across the northwest rim of China adjoining the Gobi desert. This 70-year project is designed to protect China's forests from the sandstorms of the Gobi desert.[17]
- **Inner Mongolia**: Success in the planting of desert shrubs on abandoned cropland has resulted in the stabilisation of a 7,000 hectare area, which in turn has led to the expansion of these efforts to other areas.[18]
- **Morocco**: Allocation of almost $800 million to cancel farmers' debts and convert cereal-based crops to less vulnerable olive and orchids[19] is being successfully implemented.

BUSINESS

Business has a key role to play in reducing deforestation, both in sourcing of wood products and in reducing demand.

- **Wood products sourcing and manufacture**: Sourcing of wood products is a powerful opportunity for business to play its part in encouraging sustainable forest management. Use of accredited wood supplies by businesses is an important way of driving long-term sustainable forest management. Marks and Spencer, the UK retail chain, has demonstrated important best practices in this area[20] with the details of its approach to sourcing products only from accredited suppliers being outlined in its Plan A policy.
- **Paper recycling**: Best practice is demonstrated in countries such as Germany, where over 70 per cent of paper is recycled, compared to only 30 per cent in countries such as China and Italy. If every country recycled paper at the rate that Germany does, then the amount of wood pulp used for paper production would drop by over 30 per cent.[21]
- **Logging practices**: Restricting logging activities to sustainable forests and policies of selective logging help preserve forest regions and biodiversity as well as reducing CO_2 emissions. This therefore also presents an important opportunity to address both biodiversity and Climate Change through a single initiative.

PEOPLE

People, as consumers and investors, can play an important role in the preservation of Planet Earth's forests through choices in consumption and investment.

---

16   As reported on Sustainable Development website: http://www.sustdev.org/index.php?option=com_content&task=view&id=2028&Itemid=36.
17   As reported on the BBC News website: http://news.bbc.co.uk/1/hi/world/monitoring/media_reports/1199218.stm.
18   As reported on the Sustainable Development website: http://www.sustdev.org/index.php?option=com_content&task=view&id=2028&Itemid=36.
19   As reported on the Sustainable Development website: http://www.sustdev.org/index.php?option=com_content&task=view&id=2028&Itemid=36.
20   As reported in Marks and Spencers' CSR Report: Plan A – Year 1 Review. 15 January, 2008.
21   FAO, ForesSTAT, Statistics Database, as reported at the faostst website: faostat.fao.org.

- **Consumers**: Consumers can make important choices in their purchase of sustainable wood, plant and animal products and buy from businesses with sustainability policies and practices.
- **Investors**: Investors can play a key role in ensuring they fund businesses that have clear and audited sustainability strategies.
- **Cooking fuel**: In developing countries, over half of all wood removed from forests is used as fuel for cooking. Addressing the issue of cooking fuel therefore provides an important opportunity to improve sustainability in these regions. Kenya provides an example of two important best practice approaches. Highly efficient stoves, which significantly reduce fuel consumption have been introduced with almost 900,000 already installed. The use of solar-powered stoves has also been successfully introduced as an alternative method for cooking.[22]

---

22   As reported on http://209.85.229.132/search?q=cache:l6wlAhkjBoIJ:www.martinot.info/Martinot_et_al_AR27.pdf+Kenya,+best+practice+highly+efficient+stoves+-+800,000+have+been+installed&cd=5&hl=en&ct=clnk.

CHAPTER 7

Executive Brief No. 6: Extreme Poverty

'For the first time in human history, the human race has the technologies and financial resources to eradicate extreme poverty.'

Jeffrey Sachs

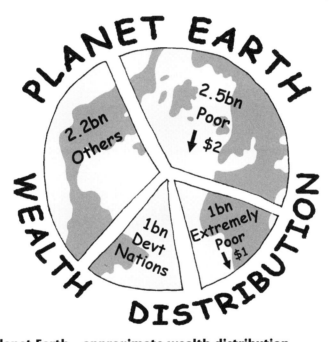

Figure 44 Planet Earth – approximate wealth distribution

SUMMARY

Despite the rapid advances in technology, agricultural methods such as the Green Revolution, financial systems and the development of the 'global village', over one billion people on Planet Earth still live in extreme poverty – surviving on less than $1 a day. Although significant aid funding is directed to this area, one of the major issues which needs to be confronted in the drive to eradicate extreme poverty by 2025 is ensuring that the required aid funds and resources actually reach those who require them. As Jeffrey Sach says, 'We have a problem with the plumbing.' Regular estimates show that less than 50 per cent of aid reaches those for whom it is intended, there is clearly too much wastage. Addressing this issue, however, provides the opportunity to rapidly increase available resources without increasing funding levels.

102 Developing a Plan for the Planet

> Over the last decade, significant progress has been made in addressing these issues and on setting global objectives and country specific strategies to eradicate extreme poverty from Planet Earth. These programmes are being coordinated through organisations such as the UN and deployed at a national and local level. They include:
>
> - Cancellation of third world debt
> - Family planning programmes
> - Education programmes
> - Healthcare programmes
> - One meal a day programmes.
>
> The results to date have been encouraging, but inconsistent across countries, dependent largely on how effectively the programmes are deployed at a local level. The priority is to increase management effectiveness to allow these approaches to be rolled out consistently across all countries.

The Current Situation

- **Extreme poverty on Planet Earth**: Over 1 billion people, almost one-sixth of all people on Planet Earth, are too hungry or destitute to get a foot on the development ladder. These are the extreme poor of the planet. Not all of them are dying today but all of these people are fighting for survival.
- **Subsistence survival**: A further 2.5 billion people live above mere subsistence. Though their daily survival is pretty much assured, they struggle to make ends meet and lack basic amenities such as safe drinking water and hygiene facilities.
- **40 per cent of the planet's population is in poverty**: In total, roughly 40 per cent of the population on Planet Earth fits into these two categories of poor or extremely poor. Regions particularly at risk are Sub-Saharan Africa and South and East Asia.
- **The 'Problem with the Plumbing'**: As already highlighted, there are issues with getting aid funding to those for whom it is intended. As Jeffery Sachs puts it, in his

Figure 45 It's not working very well is it?

book *The End of Poverty* (Penguin, 2005), 'there is a 'problem with the plumbing'. Aid funds often do not end up in the right places to have the desired affect. Of the $20 billion in aid for Afghanistan, it is estimated that very little of this actually reached the people who required it (see box insert). This is not unusual in many aid projects.

- **Chronic hunger**: In 2009, there were over 960 million undernourished people compared to approximately 920 million in 2007. Nearly two-thirds of the worlds hungry live in Asia, and Sub-Saharan Africa, where one in three people are chronically hungry.[1]
- **The poverty trap**: A large number of the extreme poor are caught in a 'poverty trap', unable to escape due to disease, physical isolation, climate stress and environmental degradation from extreme material deprivation.
- **Lack of access to resources**: Even though life-saving solutions exist to increase their chances for survival, whether through better farming techniques or bed nets that limit the transmission of Malaria, these families and their governments often lack the financial means to make crucial investments.[2] The issue becomes one of effective deployment of these solutions – not in finding new solutions.
- **Climate Change could further impact**: Climate Change could potentially drive 200 million people from their lands due to rising sea levels, desertification and flooding. This could significantly increase the numbers of extremely poor who rely on the land for subsistence farming.[3]
- **Food costs and the financial crisis**: Food shortages, increased food costs and the global financial crisis could further reduce the amount of aid and resources available to the extremely poor.
- **Fresh water supplies**: Decreasing fresh water supplies will decrease the ability to plant crops, thus increasing the number of people no longer able to grow sufficient food to provide for themselves and their families.
- **Increasing population**: As we have seen, the main population increases are forecast for the extremely poor and poor countries, with the clear result that should the issue of unsustainable population growth not be addressed in these countries, the proportion of extremely poor and poor will continue to increase as population outstrips the ability of the land and countries' resources to support them. In Sub-Sahara Africa, for example, food production is increasing at 2 per cent per annum, however, population is increasing 1.5 times faster at 3 per cent.

Solutions and Best Practice

- **Continuing to make poverty reduction a priority**: The damaging impact of the financial crisis in 2009 could potentially impact provision of aid. Despite the credit crunch, reducing extreme poverty needs to remain a priority in the Global Challenges as extreme poverty leads to high population growth, conflict and regional

---

1   As reported on the FAO website: http://www.fao.org/news/story/en/item/8836/.
2   As reported in *The End of Poverty: How We Can Make it Happen*, Sachs, J., p. 19, Penguin Group, Penguin Books Ltd, 80 Strand, London, WC28 08I, UK.
3   As reported in *The Atlas of Climate Change*, Dow, K. and Downing, T. E., p. 62, Earthscan, Dunstan House, 14a St. Cross Street, London, EC1N 8XA.

instability and therefore remains an important priority in achieving a sustainable Planet Earth.
- **On track to halve extreme poverty**: The good news is that the developing world is still on track to meet the Millennium Development Goal to halve extreme poverty from 1990 levels by 2015, and to have eradicated extreme poverty by 2025.[4] World Bank estimates in 2008 reveal 1.5 billion people in the developing world (one in four) living on less than the revised international poverty line of $1.25 a day, down from 1.9 billion (one in two) in 1981.
- **Main focus areas**: In addressing extreme poverty, it is important to identify the five key areas which require action:
  - population growth
  - education
  - health
  - debt cancellation and
  - 'fixing the leaky pipes'.
- **Assistance through debt cancellation**: This has been an important initiative. Some poor countries that have benefited from debt cancellation have been able to more than double the total sum that they invest in fighting poverty. Many countries are proactively tackling the challenge and have developed highly effective quality Poverty Reduction Programmes. Countries such as Ghana, Ethiopia, Kenya, Senegal and Uganda are good examples of this best practice.
- **Focusing resources**: Globally, there are sufficient resources and financial capabilities to reduce extreme poverty. The problem is the lack of effective management, political will and efficient deployment of these resources across and within all the countries which require this assistance.

Opportunities

- **Addressing hunger**: The Rome Declaration on World Food Security and the World Food Summit Plan of Action adopted by the World Food Summit in 1996 laid the foundation for diverse paths to a common objective – food security at the individual, household, national, regional and global levels.
  - Ensuring that food, agricultural trade and overall trade policies are conducive to fostering food security for all through a fair and just world trade system.
  - Continue to give priority to small farmers, and supporting their efforts to promote environmental awareness and low cost simple technologies.

Achieving these objectives and broad strategies is however, dependent on a clear focus, and translating these into practical applications through the three key agents of change – Government, Business and People.

---

[4] As reported on www.un.org/millenniumgoals/2008highlevel/pdf/commiting.pdf, World Bank Research Newsletter July/August, 2008.

FOCUS

The focus of extreme poverty strategies is on the 40 poorest countries shown in the following box.

> **THE WORLD'S 40 POOREST COUNTRIES**[a]
>
> Afghanistan, Angola, Bangladesh, Benin, Bhutan, Burkin Faso, Burundi, Cambodia, Cape Verde, Central African Republic, Chad, Comoros, Democratic Republic of Congo, Djibouti, Equatorial Guinea, Eritrea, Ethiopia, Gambia, Guinea, Guinea-Bissau, Haiti, Kiribati, Laos, Lesotho, Liberia, Madagascar, Malawi, Senegal, Sierra Leone, Solomon Islands, Somalia, Sudan, East Timor, Togo, Tuvalu, Uganda, Tanzania, Vanuatu, Yemen, Zambia.
>
> ---
> [a] As reported on www.factmonster.com/ipka/A0908763.html.

The Role of Government, Business and People

GOVERNMENT

Government at a national and international level area is already playing a key role in the reduction of extreme poverty in several ways:

- **Focused aid programmes**: Within each country, there are four major focus areas which can address extreme poverty.
    - **Reproductive health and family planning**: Stabilisation of population growth is critical. In Sub-Saharan Africa, the population has been growing at 3 per cent per year. However, food production has only been growing at 2 per cent per year with the result that the countries are unable to support their growing population with local resources.[5]
    - **Provision of global primary education and eradication of adult illiteracy**: Basic education provides a key opportunity for individuals and developing countries to escape the poverty cycle. These education programmes also contribute to strategies on population stabilisation as well as preventative healthcare.
    - **Provision of 'one essential meal a day for children' in targeted countries**: Basic hunger is still a major issue for over 25 per cent of the population. One of the best practice approaches to this can be the provision of school lunches for children. The logistics of this are simplified as children already are grouped at a common location and the benefits are twofold. Children are provided with basic nutrition, improving productivity, and there is also an incentive to attend school and therefore increase education levels which have been linked to both better health and population stabilisation. This provides an important quick win opportunity.

---
5   As reported in http://www.africa.upenn.edu/ECA/Food2.html.

- **Provision of basic healthcare**: Basic healthcare is critical in avoiding the poverty trap and is therefore another of the important strategies in eradicating extreme poverty. This approach concentrates on the areas that can make the best contribution such as the provision of assistance to preschool children and pregnant women. It can also be combined with family planning and contraceptive programmes such as the provision of condoms which also contributes to the eradication of HIV/AIDS. Interconnectivity can work therefore very positively when the strategies are effectively planned and coordinated on a global scale.
- **Debt cancellation**: Just as there is a poverty cycle in individuals, there is also a poverty cycle at a country level, which has been made worse by the debt of these countries to organisations such as the World Bank, International Monetary Fund, the African Development Bank, Asian Development Bank and the South American Development Bank. It is currently estimated that almost 40 countries require assistance with debt cancellation to help them break the cycle of poverty. Approximately half of these countries already are being helped, as a result of the 2005 G8 meeting in Gleneagles. There is clearly the opportunity to extend this debt cancellation programme to all countries.
- **Farm subsidies and competitive equality**: Many developing countries are being disadvantaged by farm subsidies policies. Almost $280 billion is provided in farm subsidies in developed countries, exceeding by four times the amount provided in aid for the developing countries.[6] A gradual reduction in the developed countries' farm subsidy programmes would allow developing countries to compete more effectively on the global markets in areas where they are competitive, particularly high labour-intensive agriculture. This would strengthen the developing countries' economies and promote employment. It would be an important step in developing a more balanced and equitable business approach to global development. Success in this area has been limited, with the disappointing recent DOHA Global Trade meeting in 2008. However, it is important that efforts to find common ground, particularly since free trade is a key element in dealing with the global financial crisis. Efforts to address this are continuing.
- **Fixing the leaky pipes**: The high levels of 'leakage' which diverts funds away from their targeted objectives needs to be addressed as a matter of urgency. The responsibility lies at both a national and international level.
    - The prime responsibility to identify the required target areas and funding requirements for aid lies with the country receiving the aid – the Poverty Reduction Programme. Within these plans there needs to be clear guidelines and a genuine commitment to ensure that the 'leaky pipes', mainly corruption, are fixed.
    - Organisations such as the World Bank and UN agencies are then responsible for ensuring these plans are adhered to and measuring progress and providing advice.

Clear cooperation and coordination between these international and national governments is required if these leaky pipes are to be effectively addressed. The upside is potential doubling of the available resource to the points that they are required.

---

6   As reported in Brown, L. *Plan B 3.0*, p. 146. Earth Policy Institute, W. W. Norton & Company. 500 Fifth Avenue, N.Y. 10110.

BUSINESS

Business, particularly multinational companies, have a critical role in reducing extreme poverty.

- **Investment in countries**: Well-placed investment by major corporations can help developing countries build their economies and move them out of the poverty cycle. Success in this area has been seen in many countries across both Asia and African nations.
- **Investing in people**: Global companies are taking an active part in investing not only in commercial projects but also social and community infrastructure in the countries in which they are operating. Many developing countries provide an important and cost-effective environment for these companies to produce their goods and services. Companies are increasingly recognising their corporate responsibility and investing back in the communities in which they operate.

PEOPLE

People in developed countries, as consumers and investors, can play an influential role in supporting strategies to address the Global Challenge of extreme poverty.

- **Consumers**: People can make informed choices regarding the products they purchase.
- **Investors**: Investors can take a company's social responsibility record into consideration before an investment is made – encouraging sustainability principles to be developed and adhered to by these companies.
- **Philanthropy**: Individual philanthropy can make a significant contribution to aid projects if resources end up with those for whom they were intended. It is important, therefore, to ensure these organisations have an effective audit process in place and can demonstrate effective deployment of resources.

CHAPTER 8

Executive Brief No. 7: Global Health

'We share the same planet, the same earth, the same sky, the same hopes. The things that link us together outnumber all the things that separate us.'
                                                                                      Brahma Kumaris

Figure 46 Let's not breathe today honey, let's drive. It's safer!

SUMMARY

Causes and the solutions for many of the major diseases and illness globally are known. Access to clean drinking water and good sanitation, for instance, would positively improve the health of over 1 billion people. Similarly, child and maternity health are other areas where significant improvement can be made through known treatments.

Understanding of and access to preventative treatments for illnesses such as HIV/AIDS, Malaria and Tuberculosis can positively improve the lives of millions more. These and other diseases can be managed through the effective deployment of basic healthcare facilities and immunisation programmes, particularly in low-income countries. The key to addressing global health issues is therefore less about finding solutions than about successfully deploying and managing these solutions across the countries that require them.

> In western society, lifestyle changes are required to address problems such as heart disease, alcohol, drugs, tobacco and obesity which have major impacts on human health. The challenge is not in understanding but in the political will and management effectiveness, to implement solutions.
>
> We therefore have the opportunity to manage the global deployment of actions to solve these problems. The other opportunity is to understand and address the interconnectivity between the issues of health and the other Global Challenges being faced on Planet Earth including Climate Change, access to clean water supplies and food, population and migration impacts. Many of these are in the UN Millennium Goals which provide an effective framework for building comprehensive strategies for achieving Global Health and are therefore key elements in a Plan for the Planet.

The Current Situation

- **Major causes**: The major causes of disease and death amongst the human race are unsafe drinking water and sanitation, the major infectious diseases – HIV/AIDS, respiratory diseases, Malaria, Tuberculosis and Measles, and illness associated with twenty-first century living through pollution, addictions and lifestyle. The solutions to addressing these are well known, however, not yet effectively deployed.
- **Global hazards**: The WHO also points out that large-scale and global hazards to human health include Climate Change, ozone depletion, pollution, loss of biodiversity (that is, plants and animals that provide support to human requirements such as the pollination by bees), changes in water distribution due to Climate Change and other factors impacting the supplies of fresh water, land degradation and impact on food production.
- **A new perspective on Global Health**: According to the WHO, an appreciation of the scale and influence on human health requires a new perspective which focuses on ecosystems and on the recognition that the foundations of long-term good health in populations rely on the continued stability and functioning of the biosphere's life support systems.
- **The challenge of interconnectivity**: In considering global health on Planet Earth all the factors contributing to health must be taken into consideration. This WHO diagram highlights the complex nature of the interconnectivity of human health with other Global Challenges.
- **The impact of safe drinking water**: It is estimated that over 1 billion people on Planet Earth do not have access to safe and reliable supplies of drinking water.[1]
- **The impact of maternal health**: Every week, an estimated 10,000 women die from treatable complications of pregnancy and childbirth.
- **The impact of basic sanitation**: Over 2.5 billion people – 40 per cent of the world's population – live without toilets or latrines and are unable to practice basic hygiene such as washing their hands with soap and safe water.[2] Every day 5,000

---

1 As reported in *The Atlas of Water*, Clarke, R. and King, J., p. 48, Produced for Earthscan by Myriad Editions Ltd., 59 Landsdown Place, Brighton, BN3 1FL, UK. Website: http://www.MyriadEditions.com.
2 As reported in *Time*, 16 June, 2008, p. 44.

Global Health

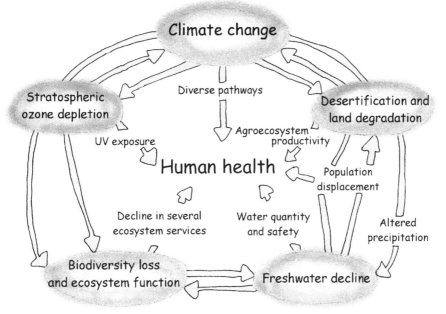

Figure 47 The interconnectivity of human health with the other key challenges

Source: World Health Organization (WHO).

children under five die from diseases associated with poor hygiene and lack of sanitation – some 1.8 million unnecessary deaths per year.[3]

- **The impact of tobacco**: It is estimated that more than 5 million people are killed by tobacco each year. This is more lives lost than through HIV/AIDS, TB, maternal mortality, motor vehicle accidents, suicide and homicide combined. If current trends continue unchecked, by 2030 tobacco will be the leading cause of death and disability, killing more than 10 million people annually, 70 per cent of them in developing countries.[4]
- **The impact of alcohol**: It is estimated that over 70 million people suffer with diagnosable alcohol use disorders. Alcoholism is the fifth leading risk factor for premature death and disability and therefore both a costly health and social problem.[5]
- **The impact of HIV/AIDS**: This is one of the key health issues being faced by the human race, particularly in the African subcontinent. HIV/AIDS affects not only the individuals concerned but it also seriously damages the social and economic infrastructure of their communities.

In just over two decades, the HIV/AIDS pandemic has claimed 20 million lives and infected 38 million people. In 1990, 10 million people were infected with the virus, however by 2004, the number had risen to 78 million, of which 38 million had died, and 39 million were living with the virus. In some areas of Sub-Saharan Africa, 25

---

3 As reported in *Time*, 16 June, 2008, p. 44.
4 As reported on the EMRO website: http://www.emro.who.int/TFI/facts.htm.
5 As reported in WHO Global Status Report on Alcohol, 2004, WHO, 20 Avenue Appa, 1211, Geneva 29, Switzerland.

per cent of the workforce is HIV positive. Young people aged between 15–24 account for half of all new HIV infections – one every 14 seconds. Despite this, fewer than 20 per cent of people at high risk of infection have access to proven prevention techniques.

- **Sub-Saharan Africa**: The majority of people living with HIV/AIDS in the world are Africans. Of the 39 million people who were living with HIV/AIDS at the end of 2004, 25 million (or over 66 per cent) were in Africa. The social and economic consequences are disastrous. As agricultural workers, particularly younger people, are infected, food production is reduced. Companies have increased costs of higher labour turnover and the reduced productivity of sick people. As the disease strikes, there are reduced numbers of teachers and educational systems are weakened. Health systems are overwhelmed by HIV/AIDS patients, and other illnesses are not properly treated. Studies show that if 15 per cent of a country's population is HIV positive, its GDP will decline by 1 per cent a year.[6] Finally, there is a tragic legacy as there will be as many as 18 million HIV/AIDS orphans by 2010. In essence, although many of the solutions to stop the spread of AIDS are known, unless these are quickly and effectively deployed, Sub-Saharan African countries face a social breakdown with severe economic consequences, where HIV/AIDS interconnects with poverty and hunger.
- **Asia**: There are approximately 5 million people living with HIV/AIDS in Asia, with the worst affected countries being India, 2.4 million; China, 700,000; Thailand, 600,000. In Asia the epidemic is focused on high-risk groups including sex workers and drug users.[7]
- **Eastern Europe and Central Asia**: The HIV/AIDS epidemic is spreading from over 600,000 living with HIV/AIDS in 2001 to 1.5 million in 2007. The Russian federation and the Baltic states are the worst affected.[8]

- **The impact of Malaria**: This is one of the leading infectious diseases, claiming more than a million lives each year, most of whom are children. In Sub-Saharan Africa, it is estimated that one in five childhood deaths are caused by Malaria. With over 275 million cases annually, there are severe economic and health costs imposed on both families and the community.[9]
- **The impact of Tuberculosis (TB)**: TB is preventable and treatable yet kills over 1.5 million people a year. It is estimated that over 2 billion people have latent TB infection, which can be activated by the HIV/AIDs infection.[10]
- **The impact of diarrhoeal diseases**: These devastate young children under five years as their bodies are weakened through rapid loss of fluids and undernourished through lack of food. In the developing countries a child has more than one occurrence of this illness each year.[11]
- **The impact of Pneumonia**: This disease kills more children than any other infectious disease and 99 per cent of these deaths are in developing countries.[12]

---

6   As reported on the UNFP website: www.unfpa.org/swp/2004/pdf/summary.pdf.
7   As reported on the AVERT website: http://www.avert.org/aroundworld.htm.
8   As reported on the AVERT website: http://www.avert.org/aroundworld.htm.
9   As reported on the WHO website: http://www.who.int/infectious-disease-report/pages/textonly.html.
10 As reported on the WHO website: http://www.who.int/infectious-disease-report/pages/textonly.html.
11 As reported on the WHO website: http://www.who.int/infectious-disease-report/pages/textonly.html.
12 As reported on the WHO website: http://www.who.int/infectious-disease-report/pages/textonly.html.

- **The impact of pollution**: Health and pollution obviously do not sit well together. A study by the University of California and Boston Medical Centre identified over 200 diseases which are linked to pollution.
  - The WHO reports that an estimated 3 million deaths a year occur from air pollution. This is three times higher than the number of road traffic deaths. In some places its now safer to drive than to breathe.
  - In 2004, 48 of the 50 states in the United States issued advisory warnings against eating fish from local lakes and streams due to their high mercury content. Research indicates that almost 20 per cent of women of childbearing age in the United States have enough mercury in their blood to harm a developing embryo.[13]

  Therefore, when managing health problems, it is essential that reducing pollution, and the impact that it is having on the health of the people, is also addressed.
- **The impact of drug addiction**: Over 200 million people between the ages of 15 and 56, or almost 5 per cent of the world population have used opiates, cocaine, cannabis or amphetamines at least once. This has a significant damaging impact on both the individuals and the communities they inhabit.
  - Only one in five (20 per cent) of addicts get treatment.[14]
  - The social and medical costs of drug addictions are immense, with clear interconnectivities such as increased vulnerability to diseases such as HIV/AIDS.
  - Other social and economic impacts are also significant with drug trafficking being a major element in criminal activities including money laundering.
- **Eyeglasses**: Some 1 billion people, mostly in developing counties, who need eyeglasses to read write, work and get about their daily lives do not have access to them. At least 10 per cent of this group is made up of children of school age.[15]
- **Health inequalities**: Healthcare quality is highly uneven. This includes access within countries, and the discrepancy of access between countries, whether measured by the breadth of medical facilities or treatment coverage.

Opportunities and Best Practice

OPPORTUNITIES

Significant progress is being made in the area of Global Health through focused programmes such as the Millennium Development Goals. Every year, 3 million more children under five years survive, 2 million lives are saved by immunisation and 2 more million people now receive AIDS treatment compared to the period prior to the establishment of the goals.[16]

---

13 As reported in Brown, L., *Plan B 3.0*, p. 48. Earth Policy Institute, W W Norton & Company. 500 Fifth Avenue, N.Y. 10110.
14 As reported in UN Office on Drugs and Crime, 2008 Report, UNDOC, Vienna International Centre, PO Box 500, A 1400 Vienna, Austria.
15 As reported on the World Bank website: http://web.worldbank.org/WBSITE/EXTERNAL/NEWS/0,,content MDK:21860878~pagePK:64257043~piPK:437376~theSitePK:4607,00.html.
16 As reported on the UNICEF website: www.unicef.org/specialsession/about/sgreport-pdf/sgrep_adapt_part1_eng.pdf and MDG, 2008 Monitoring Rrport, p. 2.

The UN Millennium Declaration[17] outlined the following strategies that support these objectives:

- Urging the international community to support the Global AIDS and Health Fund.
- Strengthening healthcare systems and addressing factors that affect the provision of HIV/AIDS-related drugs
- Supporting and encouraging the involvement of local communities in making people aware of such diseases.
- Urging national governments to devote a higher proportion of their resources to basic social services in poorer areas since this is crucial in prevention diseases.
- Supporting other initiatives based on partnerships with the private sector and other partners in development.

The challenge with all of these objectives and solutions is to translate these global initiatives into local practice. However, progress is being made:

- **Scientific advances**: Significant advances have been made in science, which mean that the major diseases can now potentially be prevented, and in some cases cured, when the right resources are deployed. Recent examples include Polio, TB and HIV/AIDS.
- **Success of global fighting funds**: Since 2000, the Global Fund to fight HIV/AIDS, TB and malaria has distributed $8.6 billion in grants to 136 countries and secured treatments for 1.1 million people living with HIV/AIDS.[18]
- **An integrated approach**: In 2009, the US administration announced a $63 billion, six year health initiative to fight AIDS, tropical diseases and other illnesses. President Obama said, 'We cannot simply confront individual preventable illnesses in isolation.'

The world is interconnected and that demands an integrated approach to global health.

- **Increased coordination**: Significant progress has already been made in increased coordination and deployment of resources through large philanthropic funds such as the Bill and Melinda Gates Foundation and through the Millennium Goals projects.
- **Commercial progress**: Low-cost, self-correcting spectacles, invented by Josh Silver, are being made by an NGO for between $5 and $10, with an ultimate target price of about $1.[19]
- **Increased healthcare**: Healthcare interventions can significantly reduce maternal deaths but need to be made more widely available. Similarly, female education regularly leads to improved hygiene not only for women, but for the whole family – which in turn significantly improves health prospects.

---

17   As reported on the OHCHR website: http://www2.ohchr.org/English/law/millennium.htm.
18   As reported on the Oxfam website: http://www.oxfam.ie/pdfs/campaigner/2008-spring.pdf.
19   As reported on the Worldbank website: http://web.worldbank.org/WBSITE/EXTERNAL/NEWS/0,,content MDK:21860878~pagePK:64257043~piPK:437376~theSitePK:4607,00.html.

The Role of Government, Business and People

GOVERNMENT

National and international government groups clearly play an essential role in establishing and maintaining critical infrastructures such as water supplies and health services which are essential to achieving the global health objectives.

- **Safe and reliable water supplies**: The provision of water supplies that are safe and reliable is a top priority and is already making a significant impact in key areas.
  - **Waste disposal**: Water-free waste disposal systems can be used instead of the costly water-based sewerage removal and treatment systems common in developing countries.
  - **Rehydration**: In Bangladesh and Nepal mothers have been taught to use boiled water and administer a rehydration solution to treat diarrhoea at home, an example of best practice which has dramatically reduced infant and child deaths.[20]
- **Infectious diseases**: Another major opportunity to improve global health is through the eradication and spread of infectious diseases such as HIV/AIDS, Malaria and TB. Best practice in these areas is being successfully deployed in many countries.
  - **HIV/AIDS**: HIV/AIDS has had a devastating effect on many communities – particularly countries in Africa and the Indian subcontinent. However, it is now spreading to most countries. For example, new cases of HIV/AIDS in China have risen by 45 per cent in the last two years and it is now the leading cause of death among infectious diseases on the mainland. However, best practices are being developed and deployed successfully to address this challenge.
  - Education programmes: Implementation of effective education programmes is an important best practice for the HIV/AIDS high-risk groups.
  - **Rolling out prevention technology**: As seen from the example of Iran and from studies in Africa, changes in sexual practices require not only practical education, but also the provision of a socially acceptable context within which to prevent the spread of HIV/AIDS. For example, abstinence and the use of condoms. The first requires no technology but is not always culturally and socially acceptable. The distribution of condoms – one of the 'technologies of prevention' therefore becomes even more important. It is currently estimated that of the approximately 18 billion condoms required, only 2 billion are being distributed today. Costs of getting the remaining 16 billion condoms to the required groups are estimated to be approximately $3 billion per annum.[21]
  - **Providing treatment**: Providing treatment for HIV/AIDS encourages people to get tested for the disease. It also increases awareness and understanding so people are less likely to pass the disease on to others. Best practices such as early treatment increases the life expectancy of these sick people and reduces the devastating economic and social damage caused by HIV/AIDS. However, the cultural problems in providing treatment are significant and are illustrated in the case of Botswana

---

20 As reported in the Rehydrate website: http://rehydrate.org/dd/dd52.htm.
21 As reported in Brown, L., *Plan B 3.0*, p. 145. Earth Policy Institute, W W Norton & Company. 500 Fifth Avenue, N.Y. 10110. As reported in *The Atlas of Water*, Clarke, R. and King, J. Produced for Earthscan by Myriad Editions Ltd., 59 Landsdown Place, Brighton, BN3 1FL, UK. Website: http://www.MyriadEditions.com. p. 48.

(see box insert). The challenge is to provide appropriate medical treatment to the 5 million infected people in territories such as Africa. Fortunately, this initiative is already well underway through the '3 by 5' initiative. This was launched by UNAIDS and WHO in 2003, with a global target of providing 3 million people living with HIV/AIDS in low- and middle-income countries with life-prolonging antiretroviral treatment[22] but funds must be increased if the objective is to be achieved.

- **Malaria, Polio, Smallpox**: Best practice is to be found in many areas in the provision of basic vaccination programs to prevent these diseases. There are already a number of important initiatives by groups such as the WHO, the Bill and Melinda Gates Foundation, the Ted Turner/UN Foundation, UNICEF and Rotary International. The challenge is to research, develop and then provide new vaccines on a scale which substantially eradicates these diseases. When well coordinated and implemented on a global scale, this best practice can be highly effective, as demonstrated by the Rotary initiative on Polio. Smallpox has been eradicated globally through a sustained and integrated programme (see box insert). Since the 1940s the number of Malaria-free countries has doubled.
- **Mosquito netting**: Another opportunity for reducing Malaria is the distribution of treated mosquito nets in vulnerable countries.
- **TB**: In Lima, Peru, the US Centre for the Promotion of Health (CPH) and their Peruvian partner Socios de Salma introduced a TB programme. Three out of four participants were cured. The WHO has therefore now recommended this best practice programme for global roll out.
- **Tobacco-free initiatives**: A major cost to global medical systems and the cause of illness and disease is the use of tobacco, mainly through smoking cigarettes. WHO studies estimate that in 2005, over 5 million people died of smoking-related diseases, which include heart disease, stroke, respiratory illness and some forms of cancer.[23] The six WHO MPOWER strategies are:
  - Monitor tobacco use and prevention policies.
  - Protect people from tobacco smoke.
  - Offer help to quit tobacco use.
  - Warn about the dangers of tobacco.
  - Enforce bans on tobacco advertising, promotion and sponsorship.
  - Raise taxes on tobacco.

Increased leadership and support is coming from the WHO Tobacco Free Initiative, and philanthropic institutions such as the Bloomberg Philanthropies and the Bill and Melinda Gates Foundation are playing an increasingly key role.[24]

- **Provision of basic healthcare services**: The success in dealing with all of these major health problems, particularly in developing nations, is through the provision of properly staffed and equipped local health clinics, and services such as education and vaccinations as well as through treatment.

---

22 As reported on the WHO website: http://www.who.int/3by5/en/.
23 As reported on the WHO website: http://www.who.int/mediacentre/news/releases/2008/pr04/en/index.html.
24 As reported on the WHO website: http://www.who.int/mediacentre/news/releases/2008/pr04/en/index.html.

BUSINESS

Business can and does play an important role in contributing to the achievement of global health. Many companies already provide health services to their employees.

- **Provision of services**: Businesses in developing countries often provide comprehensive clinics, staff welfare and community support programmes as well as providing healthcare funds.
- **Development and distribution of treatment**: Many of the large global healthcare companies provide philanthropic services and medication supplies free of charge or at low cost in developing countries.

PEOPLE

People have an important role to play in modifying their own individual behaviour and participating in good health education programmes.

- **A tobacco-free environment**: The damage to health from smoking is widely acknowledged in countries and legislation to create smoke-free zones is widespread in developed countries. However, every individual has the opportunity to take responsibility for their own health. There is also the opportunity to play an active role in lobbying government to ensure that passive impacts of smoking are effectively controlled.
- **Lifestyle**: Knowledge about health lifestyle, exercise, weight control, healthy eating and relaxation is now widely available in many countries. However, it is still the responsibility of each individual to understand and implement these healthy lifestyle strategies.
- **Support for health programmes in developing countries**: Many individuals already contribute to health aid programmes for developing countries through charitable contributions. Critical reviews of such programmes are essential to ensure that the aid donations are getting to the targeted recipients and not being wasted on high administration costs.
- **Aid management**: A due diligence process increases the accountability of groups delivering health aid and improves the quality of services.

HIV/AIDS – THE CULTURAL CHALLENGE

Jonny Steinberg had the idea for writing his book *Three Letter Plague* while reading Edwin Cameron's book *Witness to AIDS* with an account of an experiment in managing HIV/AIDS in Botswana.

Knowing that up to a third of its population had HIV/AIDS, and that about 100,000 people were in urgent need of the required drugs for treatment, the Government of Botswana announced that it would offer free antiretroviral treatment to every citizen with HIV/AIDS. It was a dramatic declaration of intent, unprecedented in Sub-Saharan Africa. By the time the drugs had hit the shelves and health personnel were ready to administer treatment, just about every person in Botswana knew about it.

And yet, on the last day of 2003, more than two years after the launch of the programme, only about 15,000 people had come forward for treatment. The rest, over 85,000 people, had stayed at home. The majority would now be dead.

Why did they not go and get the drugs?

'Stigma' is Cameron's answer: 'people are too scared – too ashamed – to come forward and claim what their government is now affording them – the right to stay alive – in some horrifically constrained sense they are "choosing" to die rather, rather than face the stigma of HIV/AIDS and find treatment.'

Source: Three Letter Plague: A Young Mans Journey Through a Great Epidemic, Steinberg, J., Published by Vintage. Reprinted by permission of the Random House Group Ltd.

CHAPTER 9

Executive Brief No. 8: Universal Education

'Education is the most powerful weapon you can use to change the world.'
Nelson Mandela

'The surest way to keep a people down is to educate the men, and neglect the women. If you educate a man, you simply educate an individual, but if you educate a woman, you educate a family.'
Dr J. E. Kwegyir Aggrey

Figure 48   Give a person a fish and they eat for a day..., Teach a person to fish, and they eat for life ..., Teach a person to read and they may end up running a global food supply company providing food to millions, funding reading programmes in other developing counties and sustainability programmes for all of Planet Earth

SUMMARY

Global Education can dramatically improve our response to the key challenges such as population stabilisation, extreme poverty and Climate Change – and also strengthen and enhance the lives of the individuals concerned and the countries to which they belong.

For this reason, there has been a strong focus on Global Education as part of the UN Millennium Goals. The main priorities are increasing education levels for primary school children, eliminating gender disparity and addressing illiteracy.

120 Developing a Plan for the Planet

> Key solutions are being developed to address these goals including the World Bank Education for All Programmes, school lunch programmes which tackle both the issues of education and malnutrition and specifically targeted female education programmes.
>
> Success to date has been mixed with uncertainty at this stage as to whether the Millennium Goals for 2015 will be achieved across all countries. There are still about 100 million primary school children out of school worldwide. Therefore universal education remains a very high priority for the international community.

The Current Situation

- **Basic literacy skills are essential**: Worldwide, over 750 million adults lack basic literacy skills. Approximately 64 per cent of them are women.[1] Illiteracy reduces the spread of new knowledge, for example, on modern agricultural methods and building up of a skilled workforce and it inhibits the empowerment of women.
- **Education is a key building block in addressing the Global Challenges**: The provision of good basic education and specialist knowledge and skills is the foundation on which all programmes to meet the Global Challenges are built. Each year of schooling raises earning power by 10 to 20 per cent. Literacy further contributes to improved hygiene and therefore overall health.

Figure 49   I think there are a few pieces missing ...

---

1   As reported on the UNESCO website: http://www.unesco.org/education/gmr2008/highlights-en.pdf.

- **Impact of illiteracy**: Half the world's population is under 25 and in developing countries 85 per cent of the population is aged between 15 and 24. Nearly half of young people are unemployed, which in part can be attributed to the fact that 113 million young people are illiterate. In Sub-Saharan Africa, less than 20 per cent of young people complete secondary school.[2]
- **Impact of population growth**: Illiteracy and population growth reinforce one another as illiterate women have larger families than literate women. In Brazil, for example, illiterate women have an average of six children and literate women have only two. Worldwide only 68 adult women are considered literate for every 100 men.[3]
- **Impact on women**: Women's earnings are lower than men's, and girls take on more responsibility than boys for child and elderly care, collecting food and water, and therefore have less opportunity for education.
- **Impact on health**: Lack of basic education and health issues are also interconnected. The poor and uneducated often do not understand the mechanisms of disease transfer and do not take action to protect themselves and are not informed of the preventions and cures which are available to them. The positive side of this is that where female education takes place on hygiene, these methods are transferred to the whole family.
- **Inadequate facilities**: Attendance is higher in urban than in rural areas but being poor is the main determinant. Crowded and dilapidated classrooms, too few text books and insufficient instruction time are widespread in many developing countries and fragile states.[4]
- **Lack of teachers**: There is a lack of primary teachers worldwide. 18 million new primary school teachers are needed to achieve universal primary education by 2015.[5]
- **Impact of tertiary education**: In addition to basic education, the number of people in a country who go on to higher education – usually referred to as tertiary graduates – is becoming more important in a knowledge-based world.
- **Access to primary education**: In Oceania, almost two-thirds of children of secondary school age are out of school. In Sub-Saharan Africa the figure is 25 per cent. Refugee children are often denied education opportunities.

    However, simple statistics about the number of children attending school can be misleading. In many developing counties, less than 60 per cent of primary school pupils who enrol in first grade reach the last grade. Teacher ratios of 40:1 are common and many teachers are inadequately qualified. Shortage of books, paper, pencils and other basic education tools also reduce the quality of education.[6] Improving the quantity and quality of education calls for a systems view since improving one aspect in isolation does not produce good results.
- **Lack of equity**: Resources are often skewed in favour those with the greatest wealth rather than those with the greatest need. The marginalised such as slum dwellers, child labourers and the extremely poor are disadvantaged.

---

2   As reported on the WorldBank website: web.worldbank.org/WBSITE/.../NEWS/0,,content MDK:21869960~pagePK:6425704 3~piPK:437376~theSitePK:4607,00.html World Bank Human Development Network. Education. 2008.
3   As reported on Earth Policy website: http://www.earth-policy.org/Books/PlanB_ch5_socialdivide.pdf.
4   As reported on the UNESCO website: http://www.unesco.org/education/gmr2008/highlights-en.pdf.
5   As reported on the UNESCO website http://www.unesco.org/education/gmr2008/highlights-en.pdf.
6   As reported on the MDGS website: mdgs.un.org/unsd/mdg/Resources/Static/Products/Progress2008/MDG_Report_2008_En.pdf and Human Development Network – Education. World Bank, 2008.

- **Funding requirements**: The amount of money which is needed globally to make significant further improvements in education is enormous, and at present inadequate, because the benefits are not fully understood. The education budget of a single country like France, Germany, Italy or the UK outweighs education spending across the entire Sub-Saharan African region. The USA, home to 4 per cent of the world's children and young people, spends 28 per cent of the world's global education budget according to Global Education Digest 2007.[7]
- **Funding sources**: Funds for this challenging task come from many sources: national governments, self-financing by individuals (in many developing countries this can be as high as 10 per cent of education expenditure), charities and aid donors. The World Bank is one of the world's largest providers of external funding for education. They have a dual focus: to help countries achieve universal primary education and to help countries build the higher level and flexible skills needed to compete. Lending is only part of their contribution and is complemented by policy advice, sharing of global knowledge and good practices.
- **Research and development**: The supply of tertiary graduates also governs the level of R&D within a country. China has the most tertiary graduates in the world – 2.4 million in 2006. This is more than the top three OECD countries combined: the USA has 1.4 million; Japan has 0.6 million; France has 0.3 million.[8]

R&D expenditure has grown worldwide between 1996 and 2005 but developing countries invest less than 1 per cent of GDP in R&D. In most developed countries, R&D activities are largely financed and conducted by the business sector but the public sector plays a major role in developing countries.[9]

MEXICO'S POOREST PEOPLE HELPED BY COMPUTER AND INTERNET TRAINING

The State of Veracruz Public Education Department, Mexico, use 24 all-terrain vehicles, each equipped with computers, satellite Internet connection and trainers to teach 120,000 people in 200 communities. These communities had computers in public places through Federal and State grants but they were severely underutilised.

Programmes in this Vasconcelos Programme are tailored to meet local economic, health and educational needs and are aligned to the indigenous cultures.

The programmes have, for example, helped more than 6,500 small-scale farmers to monitor coffee prices online, together with access to information about fertiliser and loans. A small jewellery business used the programme to improve its marketing efforts with a rise of 25 per cent in the family's income.

The Bill and Melinda Gates Foundation, in 2008, awarded the Vasconcelos Programme a $1 million 'Access to Learning Award'.

---

7   As reported on the UNESCO website: http://www.uis.unesco.org/template/pdf/ged/2007/EN_web2.pdf.
8   As reported on the UNESCO website http://www.uis.unesco.org/template/pdf/wei/2007/WEIfactsheet_EN.pdf.
9   As reported in the UNESCO website www.unesco.org/science/psd/wsd07/global_perspective.pdf.

Opportunities and Best Practice

OPPORTUNITIES

- **Progress is being made**: As with all of the global strategies, it is important that there is focus on the areas where maximum impact can be achieved. The greatest need is obviously in developing countries. The World Bank started supporting education in developing countries in 1963 and to date has transferred more than $41 billion in loans and credits for education. In the last five years lending has averaged about $2 billion annually.[10] And progress is being made. The proportion of children in developing countries who have completed primary education rose from 79 per cent in 1999 to 85 per cent in 2006.[11]
- **Increasing opportunity**: There is an opportunity for more investment in training people in appropriate local skills once they have achieved basic literacy and numeracy. Focused skill training has however often taken second place to general education.
- **The power of computers**: There is great scope for harnessing the learning power of computers and the Internet. For example, Mexico's poorest people are being helped by computer and Internet training (see box insert). Microsoft in 2004 started its 'ICT in Schools Africa Pathfinder Project' to test the feasibility of using refurbished computers to help technologically disadvantaged communities and to build capacity for economic and educational development. So far Microsoft, with partners, has sent over 30,000 computers to Africa. Computers4Africa are the project managers. Other organisations are providing computers and training. Computers4Africa estimate that every computer can change the lives of 17 students and their families.
- **Progress reducing gender disparity**: Women are still disadvantaged in terms of opportunities to attend school and further education. The goal of eliminating gender disparity in global primary and secondary education appears to be achievable by 2015, with the exception of Sub-Saharan Africa.

The Role of Government, Business and People

GOVERNMENT

Both international and national governments again play key roles in achieving global education objectives and there are a number of best practices that can be built upon.

- **Funding education programmes**: The World Bank is continuing to support best practice by taking a lead in with its 'Education for All' programmes. The objective is for all children in developing countries to be getting primary education by 2015.[12]
- **School lunch programmes**: School lunch programmes are a very effective incentive to get children to attend school. Academic performance increases because

---

10 As reported on the World Bank website: siteresources.worldbank.org/INTHOUSINGLAND/Resources/339552-1153163100518/Thirty_Years_Shelter_Lending.pdf and Human Development Network – Education. World Bank 2008.

11 As reported on the BMZ website: www.bmz.de/en/figures/millenniumsentwicklungsziele/mdg2.html.

12 As reported on the UNESCO website: http://portal.unesco.org/education/en/ev.php-URL_ID=46881&URL_DO=DO_TOPIC&URL_SECTION=201.html.

poor nutrition is linked to diminished productivity. In addition, the attraction of meals at schools increases the overall length of time spent in education.
- **Focus on female education**: Lawrence Summers, President Obama's Economic Adviser, stated that, 'Educating women yields a higher rate of return than any other investment in the developing world.' The World Bank's study of 100 counties found that a 1 per cent increase in the level of women's education generates an additional 0.3 per cent in economic growth. A further study of 63 countries showed that women's education accounted for 43 per cent of all progress in reducing child malnutrition.

   There are many examples of best practice which can be drawn upon. For example, Iran has focused specific education programmes around female education. Ethiopia is using Girls' Advisory Committees. Parents seeking early marriage for their daughters are encouraged by these groups to keep the girls at school. Brazil and Bangladesh provide further examples of best practice strategies by providing small scholarships for girls or funding for parents to help poor families get basic primary education.[13] Scholarships are offered in exchange for a commitment to teach for at least five years following graduation. In Australia, teaching graduates are able to fast track their career progress by spending time teaching disadvantaged groups, particularly in isolated locations.
- **Fill the gaps in aid programmes**: Despite this progress, there are gaps in the funding requirements. The donor pledges to double aid to halve poverty made at Gleaneagles in 2005 have not been fulfilled, leading to financing gaps in educational budgets. Moreover, aid to basic education is heavily concentrated on the UK, Netherlands and the International Development Association and seen as a low priority by many donors. It is essential to fulfil those promises and to widen the donor base of aid for basic education.

BUSINESS

Business has an important role to play, and there are good examples of business developing and providing educational infrastructure and services. This is outlined more fully in Chapter 5, Driving Change and Making it Happen, however a number of initiatives are outlined below.

- **Employee education programmes**: Many businesses provide employee education and skills programmes in developing countries to improve overall productivity and employee satisfaction. These trained people become an important asset to their communities as well as the business.
- **Community education programmes**: Many businesses, as part of their corporate responsibility strategies, provide resources and facilities to local communities to develop their education capabilities.

It is always important that new facilities can be supported by the local infrastructure and other complementary services. There are many examples of schools being built without the long-term provision of funds for teachers and books. A holistic approach is essential for success.

---

13   As reported on the Earth Policy website: www.earth-policy.org/Books/PlanB_ch10_socialresponse.pdf, p. 135.

PEOPLE

Individuals can also contribute significantly in education programmes, though both physical and financial contributions.

- **Funding and resourcing**: Philanthropic funding by individuals is a valuable resource. This may be in financial terms, or in person, with many young people taking 'time out' to spend a year teaching in developing countries. Some companies, as part of their social contribution, give employees time off to help in such projects.

CHAPTER 10
Executive Brief No. 9: Conflict and Peace

'It does not take special qualities in political leaders to start a war, but it takes a special type of leader who can blend vision with pragmatism to lead the way from war to peace.'

Dan Smith

Cold War

Figure 50 **Mutual assured destruction syndrome (MAD)**

SUMMARY

Peace has been an elusive quality for the human race. With two major world wars, and numerous regional and national conflicts within the last century, the task of achieving peace has been a continuing challenge. Now, however, with increasing understanding of the urgency with which the key Global Challenges need to be addressed by the human race – it is increasingly acknowledged that that key finances and resources required to address the planet's sustainability are available and focused on achieving this goal. In this crisis period, it is clear the cost of conflict and war far exceeds the cost of peace and long-term sustainability.

The threat of the ten Global Challenges demand a period of peace in which the human race and the global economy can move to a new green economy. As we have seen, challenges such as Climate Change, food and water resources, biodiversity and energy supplies demand answers within the next decade. Time is not on our side – a clear focus of efforts and resources is therefore critical for success.

The good news is that the mechanisms and the processes for averting, addressing and winding down conflict are now better understood than ever before. The challenge is to apply the knowledge and understanding developed over the last century, through the efforts of the UN, and the many peacekeeping groups operating on a global scale, to obtain this sustained period of peace.

This may seem to be unrealistic when one looks at current levels of war and conflict – until it is compared to the alternative. It is only when we as a human race fully understand that the consequences of ignoring the major Global Challenges literally could mean the end of human civilisation as we know it – that political will for peace will prevail.

The Current Situation

- **Military spending versus peacekeeping**: In 2007, the estimated global military expenditure was over $1,200 billion, approximately 2 per cent of global GDP.[1] This is compared to the $7 billion spent by the UN on international peacekeeping, and the annual International Development Aid spending of $100 billion.[2]
- **Impact**: Conflict and war causes direct human suffering and, as seen, diverts significant financial and human resources which could be focused on the urgent human priorities of dealing with Climate Change, development of alternative energy sources, food and water management, better health and education, and reducing poverty. The persistence and scale of human conflict is unfortunate considering the advances which have been made in so many other areas of technology, society and economics.
- **Global conflict decreasing**: The good news is that over the last 50 years or so, major wars between the great powers have dramatically reduced compared with the previous 50 years.
- **Regional conflicts increasing**: The concern is that civil wars have become more common, longer lasting and more difficult to resolve. It has been estimated that typically a civil war reduces a country's growth by about 2 per cent and on average costs $62 billion.[3] This clearly has implications for the developing countries where many of these conflicts are taking place – holding back development and the ability to address the Global Challenges.

    These averages disguise wide variations but are indicative of the scale of the economic damage. Impossible to measure is the human suffering, the inevitable collapse of good governance and the long-term legacy of bitterness and mistrust between warring factions.
- **Military spending versus other spending**: Over half of the US 2007 tax payer's dollars were allocated to military and related spending.

---

1   As reported on the Global Issues website: http://www.globalissues.org/article/75/world-military-spending.
2   As reported on the Wikipedia website: http://en.wikipedia.org/wiki/Development_aid.
3   As reported in *Global Crises, Global Solutions*, Lomborg, B, p. 135, Cambridge University Press, The Edinburgh Building, Cambridge, CB2 2BU, 2004.

Figure 51 **The real costs of war!**

- **Reallocating resources**: The following table provides a breakdown of the 2006 estimates of current arms expenditure. As already highlighted, total expenditure increased to over $1200 billion in 2007.

Table 6 **Largest military expenditures, 2006**[a]

Rank	Country	Spending level ($ billion)	Per capita ($)	World share (%)
1.	United States	$528.7	$1,756	46%
2.	United Kingdom	59.2	990	5
3.	France	53.1	875	5
4.	China	49.5	37	4
5.	Japan	43.7	341	4
6.	Germany	37.0	447	3
7.	Russia	34.7	244	3
8.	Italy	29.9	514	3
9.	Saudi Arabia	29.0	1,152	3
10.	India	23.9	21	2
11.	South Korea	21.9	455	2
12.	Australia	13.8	676	1
13.	Canada	13.5	414	1
14.	Brazil	13.4	71	1
15.	Spain	12.3	284	1
	Subtotal, top 15	**963.7**		**83**
	World	**1,158**	**177**	**100**

a   Market Exchange Rate. 2. Data for Iran and Saudi Arabia include expenditure for public order and safety and might be slight overestimates.3. The populations of Australia, Canada, and Saudi Arabia each constitute less than 0.5 per cent of the total world population. *Source:* SIPRI Yearbook 2007, Stockholm International Peace Research Institute.

- **Post-Cold War**: Although many people hoped that the end of the Cold War in 1989 would mark a new era of peace there have been over 120 conflicts worldwide since then.[4]

    Further, although military spending dropped following the end of the Cold War, it has been increasing constantly since 1998, and is now returning to pre-Cold War levels.[5] In 2007, there were 18 significant armed conflicts taking place.

Table 7   Significant ongoing armed conflicts, 2007[a]

Main warring parties	Year began
Middle East	
U.S. and UK vs. Iraq	2003
Israel vs. Palestinians	1948
Asia	
Afghanistan: U.S., UK, and Coalition Forces vs. al-Qaeda and Taliban	2001
India vs. Kashmiri separatist groups/Pakistan	1948
India vs. Assam insurgents (various)	1979
Indonesia vs. Papua (Irian Jaya) separatists	1969
Philippines vs. Mindanaoan separatists (MILF/ASG)	1971
Sri Lanka vs. Tamil Eelan (conflict ended May 2009)	1978
Africa	
Algeria vs. Armed Islamic Group (GIA)	1991
Côte d'Ivoire vs. rebels	2002
Democratic Republic of Congo and allies vs. Rwanda, Uganda, and indigenous rebels	1997
Somalia vs. rival clans and Islamist groups	1991
Sudan vs. Darfur rebel groups	2003
Uganda vs. Lord's Resistance Army (LRA)	1986
Europe	
Russia vs. Chechen separatists	1994
Latin America	
Colombia vs. National Liberation Army (ELN)	1978
Colombia vs. Revolutionary Armed Forces of Colombia (FARC)	1978
Colombia vs. Autodefensas Unidas de Colombia (AUC)	1990

[a] Note: As of December, 2007. 1. Where multiple parties and long-standing but sporadic conflict are concerned, date of first combat deaths is given. 2. 2002 cease-fire collapsed in 2006. Published with permission PROJECT PLOUGHSHARES 57, Erb St West, Waterloo, ON N2L6CT Canada. www.ploughshares.ca.

---

[4] As reported in *The Atlas of War and Peace*, Smith, D., p. 8, Produced for Earthscan by Myriad Editions Ltd., 59 Landsdown Place, Brighton, BN3 1FL, UK. Website: http://www.MyriadEditions.com.

[5] As reported in *Global Crises, Global Solutions*, Lomborg, B., p. 135, Cambridge University Press, The Edinburgh Building, Cambridge, CB2 2BU, 2004. As reported on http://www.globalissues.org/article/75/world-military-spending.

- **Iraq**: By the end of 2009, the direct cost to US Treasury alone will be over $1 trillion. This does not include the repatriation and compensation costs for the families of the Americans killed in the war.[6]
- **Afghanistan**: Another example where the major powers have committed themselves to addressing a terrorist threat – the Taliban – through military intervention together with heavy investment in rebuilding the country's infrastructure.
- **Nuclear weapons**: The destructive power of nuclear weapons cannot be contained in either space or time. They have the potential to destroy existing civilisation as we know it and possibly significantly alter the entire ecosystem of the planet.[7]
  - Although nuclear arms have decreased since the end of the Cold War, the high concentration of people in cities means that there is now the potential to wipe out large numbers of the global population in minutes if the current arsenal of nuclear weapons were fully deployed.
  - Although only used twice, in Japan in 1945, there have been over 2,000 nuclear weapons tests since then. There remain over 9,000 nuclear warheads on Planet Earth. Independent tests in Polynesia show evidence of impact on child stillbirths and deformities following testing on Pacific Islands, and poisoning of local fish stocks following underground testing.[8]
- **Genocide**: In the 400 years between 1492 and 1900 it is estimated that 4 million human beings died from genocide. In the 100 years between 1900 and 2000 more than 25 million people died from genocide.[9]
- **Arms suppliers**: Over 80 per cent of the arms traded are by five countries – the USA (47 per cent), Russia (15 per cent), France (10 per cent), UK (7 per cent) and Germany (5 per cent).[10]
- **Small arms trade**: 500 million small arms are available worldwide worth over $5 Billion per annum – with up to 20 per cent of this trade being illicit. Over 500,000 people are killed each year by small arms – with an estimated 60 per cent in wars and conflict and 40 per cent in crimes.[11] Arms embargos have only met with limited success.
- **Potential for increased conflicts**: Conflict within a country and with its neighbours is certain to increase if the Global Challenges of population growth and effective management of resources such as water and productive land are not successful. Conflict can also be driven by a combination of disagreements on the allocation of scarce resources and deep ethnic and religious differences. The Darfur conflict is a current example. The heart of the devastating Congo conflict is the battle to secure valuable commodities such as diamonds.

---

6   As reported on the BBC News website: http://news.bbc.co.uk/1/hi/business/7304300.stm.

7   As reported in International Court of Justice, July 8, 1996. Advisory Opinion of the International Court of Justice on the legality of the Threat or Use of Nuclear Weapons, United Nations General Assembly, A/15/218. 15 October, 1996, p. 17

8   As reported in *The Atlas of War and Peace*, Smith, D., p. 26, Produced for Earthscan by Myriad Editions Ltd, 59 Landsdown Place, Brighton, BN3 1FL, UK. Website: http://www.MyriadEditions.com.

9   As reported in *The Rise and Fall of the Third Chimpanzee*, Diamond, J., p. 256, Vintage, 2002. Random House, 20 Vauxhall Bridge Road, London. SW1V 2SA.

10   As reported on http://en.wikipedia.org/wiki/Arms_industry#World.27s_largest_arms_exporters.

11   As reported in *The Atlas of War and Peace*, Smith, D., p. 30, Produced for Earthscan by Myriad Editions Ltd, 59 Landsdown Place, Brighton, BN3 1FL, UK. Website: http://www.MyriadEditions.com.

CAUSES OF WAR

It is obviously unrealistic to expect to end all wars but we must commit ourselves to reducing them. Every dollar which is not wasted in conflict can be redirected to meeting the Global Challenges. However, we can not hope to reduce wars without understanding some of their basic causes. There is rarely a single cause so we need to consider the different influences:

- **Political**: War is one of a choice of options to secure political goals. It maybe a drive for power and territorial gain such as Hitler's Germany; or cooperation between nations to fight a common enemy – the Taliban in Afghanistan. The concern might be identified as humanitarian – Kosovo – or ideological – Vietnam. The evidence suggests that wars are more often started by authoritarian rulers than by democracies. For example, in the last ten years, it is estimated that nearly half of dictatorships have been involved in civil war.
- **Economic**: Closely linked with political ambition is the use of conflict to secure economic gain. Iraq invaded Kuwait to gain oil supplies. The Congo has many countries and factions fighting for uranium and other important commodities; much of the trouble in Angola stemmed from conflict over the ownership of diamond resources.
- **Demographic and scarcity conflicts**: This is becoming a much important potential source of conflict. Water, for example, when shared by several countries where the demands of rapidly growing populations exceeds the supply can potentially lead to conflict unless effective and cooperative management approaches are put in place.
- **Poverty**: Civil wars often take place in the poorest countries. Sometimes they are driven by ethnic or political ambitions but more often is the sense of grievance as the disparity between the many poor and the few rich turns to violence. In Mahatma Gandhi's words, 'If through our wisdom we could make secure elementary human needs, there would be no need for weapons and war.'
- **Limited success of peacekeepers**: Global and regional conflict resolution bodies have been established by groups such as the UN, EU and African nations. However, their attempts to address and resolve regional and national conflict situations have had only limited success.

IMPACT OF THE GLOBAL CHALLENGES

Population growth will further increase the potential for conflicts over limited resources as resources such as oil, water, food and high-quality land become scarce. Therefore in tackling the Global Challenges, the requirement of establishing a period of peace in which these challenges can be cooperatively addressed is essential to maximise the chance of success in achieving a long-term sustainable world.

The different ways in which Climate Change can produce security risks is illustrated in the diagram opposite.

Conflict and Peace

Key factors / Conflict constellation	State constitution, political stability	Governance structures	Economic performance and distributional equity	Social stability and demographics	Geographical factors	International power distribution and interdependency
Degradation of freshwater sources	General relevance	Effective national and international water management prevents crisis	High conflict potential in DC's due to great dependence upon agricultural sector	High risk of water crises due to rising demand in conjunction with stagnating supply	High risk of local conflict in catchment areas. Risk of regional destabilization ('spillover')	High conflict potential due to disparate national interests and local needs
Decline in food production	General relevance	High conflict where land-use rights are inequitable	High conflict potential in DC's due to great dependence upon agricultural sector	High risk of food crises due to drought and population growth and density	Risk of regional destabilization ('spillover')	Major relevance of world market conditions in cases where agricultural production declines
Increase in storm and flood disasters	General relevance. Acute pressure to act compromises government legitimacy	Effective disaster risk management prevents crisis	High conflict potential in DC's Elevated conflict potential in IC's due to dependence upon complex infrastructures	High conflict potential due to high population density and weak institutions	Risk of regional destabilization ('spillover')	General relevance
Migration	General relevance	Effective migration management prevents crisis	High conflict potential in DC's	High conflict potential due to changes in or intrumentalization of ethnic composition	Elevated conflict risk due to mounting resource competition in destination country	General relevance
General relevance for all conflict constellations	Lack of stability acts as catalyst High conflict risk in periods of state transition (e.g., democratization)	Effective performance of governance functions forestalls crisis and conflict	Poverty and socio-economic disparities increase susceptibility to crisis and potential for conflict	High risk of conflict where society structures are weak	Increased potential for conflict with neighbouring countries	Divergent impacts of economic integration and world market conditions Cooperative foreign policy and action forestalls conflict

Figure 52   The impact of climate change on security risks

Opportunities and Best Practice

OPPORTUNITIES

Although there is general agreement that war of all kinds is costly and undesirable, strategies as well as resources to deal with global and regional conflict management have met with limited success.

In spite of this, after the last decade of peacekeeping, in which the UN has been playing a more prominent role following the end of the Cold War, governments and NGOs are building experience and examples of best practice where success has been achieved.

- **Conflict resolution and the peace process**: There is a range of conflict resolution processes taking place globally in major areas of conflict, usually using third-party approaches such as the Oslo process to seek resolution.

- **Mozambique**: A solid peace has been established through encouragement by the UN and transformation of insurgent forces into a parliamentary political system.[12]
- **South Africa**: A solid peace was established through the demonstration of quality political leadership by two political rivals – President De Klerk and Nelson Mandela.
- **Northern Ireland**: The ceasefire of 1994 began the peace process, followed by the Good Friday Peace Agreement – brokered jointly between the UK and Ireland. In spite of repeated crises, there is a majority agreement that it is preferable not to return to conflict.[13]

Other regional conflicts can best be described as work in progress, with best practice still to be established.
- **Israel and Palestine**: Violence and distrust remain. Efforts by the International Community continue.
- **Speed of response**: Greater global communications and transport mean that conflict situations can be more quickly monitored and responded to.
- **Proactive programmes**: Cooperative efforts to develop agreements over issues such as water usage in countries relying on the same supplies are increasingly demonstrating an ability to anticipate and avoid potential conflict situations.

The Role of Government, Business and People

GOVERNMENT

- **Lasting peace**: This can only be achieved by dealing with the fundamental issues that make a country or region vulnerable. For example:
  - **Governance**: The absence of good governance and associated issues of trustworthy legal systems can make peace fragile. Key in established lasting peace in any country is ensuring that good governance is effectively in place.
  - **Infrastructure**: Peace rarely lasts when a country remains poor and lacks the infrastructure of health systems, schools and industrial developments. The same is required for security infrastructure without which there can be a pattern of minor conflicts over scarce resources which can quickly escalate into serious conflicts. Appropriate social and security infrastructure development is therefore key in building the potential for lasting peace.
  - **Reconciliation**: Without reconciliation between groups of different ethnic backgrounds or religious affiliation it is difficult to maintain peace within a country, The Tamil Tigers fought costly, and ultimately unsuccessful, battles to seek independence. Riots and massacres in Kenya in 2008 illustrated underlying tribal tensions. Best practice in Northern Ireland, however, demonstrates that long-lasting differences can be addressed if acknowledged and consensus built regarding long-term solutions that are win-win.
- **Peacekeeping**: Another best practice in achieving peace is clearly the peacekeeping process itself. In 2008, there were over 18 UN Peacekeeping Missions taking place

---

12 As reported on http://www.un.org/News/ossg/common.htm.
13 As reported on http://en.wikipedia.org/wiki/Northern_Ireland_peace_process.

with success in assisting the rebuilding process in Central America, Mozambique, Namibia and Cambodia. Failures, however, have occurred in regions such as Sierra-Leone and Bosnia-Herzegovina.[14] The opportunity is clearly to leverage the increasing experience and best practice in effective peacekeeping into the more challenging regions.

- **Refocus defence spending to Global Challenge defence**: As identified, approximately 2 per cent of global GDP is currently allocated to military spending. One option, with the immediate and urgent threat of the Global Challenges, is to refocus some of this significant funding towards 'defence' against the Global Challenges.

    Since challenges such as Climate Change, alternative energy sources, dwindling resources, and the loss of biodiversity of planet as we know it threaten the entire human race, increased focus of military and defence resources towards these challenges makes good sense. It can be seen that less than a 20 per cent refocus of the global military spending budget would address the additional funding required to address the gap in funding for many of the Global Challenges identified by Lester Brown in *Plan B 3.0* (see Chapter 11, Financing a Sustainable World).

    Is this realistic? Following the end of the Cold War, global defence spending reduced by approximately 30 per cent. Therefore, this funding reallocation is clearly feasible.

    However, the key question is whether there is sufficient political will by countries and international organisations to make this change. The UN has a moral authority and under Chapter VII has some powers to act in the event of war. In the real world this has little impact because such decisions have to go through the Security Council where the record is of vested national interest dominating as the UN as a body only has limited powers.

    Other groups, however, such as the G8 and the G20 have an important role to play. This has been demonstrated by the swift, cooperative action taken on the global financial crisis, which would not have been contemplated six months previously. The coordinated actions of bodies such as the G20 in London in 2009 were unprecedented, and represented an important step forward. The opportunity is therefore to act in a similar globally coordinated manner to redirect military funding and build a cooperative 'defence' infrastructure and budget to successfully address the Global Challenges.

    Although this may appear idealistic, the human race and its political systems are increasingly demonstrating an ability to develop the flexibility, innovation and cooperation required to tackle issues on a global scale.

As the impacts and demands of the Global Challenges become more evident, the willingness and ability of both decision makers and the electorate to move quickly and decisively should increase. This should enable the required 'global defence' funding and cooperative action to move from idealism to reality. In the end, there is no other choice.

---

14   As reported on http://www.prospect.org/cs/articles?article=the_coming_peacekeeping_crunch.

136 Developing a Plan for the Planet

Figure 53 You know, I really think they should have thought about the other challenges like climate change when allocating defence spending!

BUSINESS

Business can also play a role in the management of conflict situations. The key roles are reinforcing the international and national government agendas. For example:

- **Physical reconstruction**: Rebuilding in post-conflict periods and provision of new goods and services can only be done in partnership with business.
- **Economic infrastructure**: Business can contribute by increasing the total wealth of country in the difficult post conflict period.
- **Small arms trade**: Companies in the arms industries can contribute significantly by adhering rigorously to licence regulations as over 40 per cent of small arms trade is currently illegal.
- **Anti-corruption**: Business, particularly global corporations, must set an important example in stamping out bribery and corruption which are endemic during and after conflict. This is not only morally responsible and ethical, but also makes good business sense.

PEOPLE

- **Local and community mediation**: There are many skilled and successful community and NGOs which offer mediation and conciliation services. In areas of Indonesia where conflict is widespread, local district authorities play an important role in arbitrating disputes.
- **Lobbying**: The Global Challenges threaten not only all people on Planet Earth – but also all generations to come. Everybody therefore needs to take responsibility to ensure these challenges are addressed with the urgency and resourcing required. This means taking individual action where this is appropriate, acting appropriately

in the business environment, and acting to gain the appropriate responses from government.

Short-term thinking is no longer an option, with the long-term implications of the Global Challenges staring each of us in the face. The electorate can, and needs to therefore, play an active part in lobbying to achieve the required period of peace and the reallocation of defence resources to address these Global Challenges.

Finding an acceptable solution to this crucial issue of conflict management, the achievement of the required period of peace, and the associated reallocation of funding could be the most difficult of all the Global Challenges. However, a period of peace and unprecedented global cooperation would provide the most important elements in the transition to the new green economy – and therefore needs to be one of the key challenges to be addressed if long-term sustainability is to be achieved.

CHAPTER 11

Executive Brief No 10: Financing a Sustainable World

'There is cost to living on this earth. We either pay now or our grandkids pay later, and I believe we should pay as we go.'

Bernie Karl

*'It is no use saying 'We are doing our best.
You have got to succeed in doing what is necessary.'*

Winston Churchill

Figure 54   'They're beginning to get away, we'd better do something'

SUMMARY

It is becoming increasingly clear. Building a sustainable world is non-negotiable. Failure to achieve this will result in the failure of human civilisation as it is currently known. The stakes are that high. Therefore, as with any successful plan, we must put in place economic and financial frameworks with which to manage the development and funding of sustainability. These range from the development of alternative energy sources, improved food and water sources to supporting biodiversity.

> There is significant progress being made. Cap and Trade systems are putting a price on carbon and building capital to develop alternative fuel sources and offset the impacts of use of carbon fuels. Environmental finance is developing as the true cost of environmental damage is being understood.
>
> Cost estimates are becoming more rigorous, with existing shortfalls and total requirements better identified. Revenue sources are available through potential diversion of existing defence spending from military defence to global sustainability defence. Focusing financial resources must maximise impact and the speed of development of these resources. However, as we have seen, the timeframes are limited and urgent.
>
> We must move quickly and flexibly to develop and implement a financial model which both maintains the equilibrium of the current approach – which is required for everything from food production funding to sustainable R&D projects – whilst fast-tracking the changes required to move from the traditional fossil fuel-based, non-sustainable economy – to the new green economy.

The Current Situation

- **The global economy**: In 2007, the current global economy was estimated to be worth over $67 trillion per annum. This can be broken down as follows.
    - $44 trillion, or approximately two-thirds of this global GDP, is created in the top 20 countries.
    - 2 per cent or over $1,200 billion of global GDP was spent on military expenditure, which was more than ten times the investment in international aid expenditure.
    - 0.2 per cent or $100 billion of this GDP was spent on international aid.
    - The total energy consumption on Planet Earth is estimated to be worth $70 billion, of which, however, currently only 1 per cent is renewable energy.
- **The poverty economy**: Because of the increases in food costs, the ability of low-income people to afford basic food items is reduced, thus increasing the prevalence of extreme poverty.[1]

Solutions and Best Practice

Achieving a sustainable world, as we have seen, is dependent on a number of key areas:

- **Understanding**: An effective understanding of the Global Challenges and their interconnectivities and the costs required to address them.
- **Effective planning**: A global, coordinated plan on how these challenges can be addressed – with clear objectives, targets, timelines and budgets.
- **Effective implementation**: Effective and cooperative global implementation.
- **Effective financing**: However, none of this will be effective without effective financing to address these challenges.

---

1   As reported on http://poverty.developmentgateway.org/Community-Content.9258+M5e94d928388.0.html.

Effective financing for a sustainable world must address the following issues:

- **Cost assessment**: Identifying the estimated costs of addressing each of the Global Challenges. This, as we will see is complex, due to both our developing understanding of both the challenges themselves and their interconnectivities.
- **Restructuring**: Establishing an equitable global trading system will help address the issues of poverty in developing countries – thus decreasing dependency, and increasing the proactivity with which developing countries can themselves address the Global Challenges without outside support.
- **Effective implementation**: Ensuring funding needs to be effectively and efficiently deployed to properly address the challenges. Due to both time and resource constraints we can no longer afford the gross inefficiencies which have traditionally plagued our aid funding.
- **Sourcing**: Identifying reliable sources of funding to effectively finance the solutions to the ten Global Challenges is key, however, as both business and government realise the new green economy is the only sustainable economy – capital and investment funding will increase. However, short term, financing the solutions to the Global Challenges will require the rapid deployment of the financial resource available. The good news is that we are already seeing a number of options emerging:
    - **Efficiencies**: As we have seen, many organisations can significantly improve efficiencies and reduce costs through the energy usage audit and review process. These programmes therefore become self-funding, providing long-term benefits not only to the environment, but also to the business itself.
    - **Rebalancing**: Although it is not always popular politically, wealthy nations have the opportunity to increase the aid/debt reduction programmes to the developing world – and ensure that these programmes are properly managed. The nature of the global village means that effectively addressing the challenges in the developing counties will also provide significant benefits to the developed countries. There is a win-win situation available.
    - **New sources of finance**: For example, from Cap and Trade systems addressing the challenges of greenhouse gas emissions. The development of this financing system not only provides incentives for business to operate in a more sustainable manner, the funding generated provides important funding for other sustainability projects and initiatives.
    - **Reduction of subsidies**: Subsidies often distort open markets and hinder developments in poor countries, for example, the European farm subsidies. Addressing these subsidies provides the opportunity to build a more equitable playing field for the developed and developing countries and in the long term lead to greater efficiencies.
    - **Innovative use of capital**: For example, micro-financing. The success of micro-financing systems in countries such as Bangladesh have demonstrated how a small amount of seed funding to small business ventures can play an important part in addressing poverty in these countries. In 2009, a gap of $1.8 billion in micro-financing funds due to the credit crisis placed 150 million micro-finance customers at risk. To address this, a new micro-finance Enhancement Facility has been set up to provide $500 million to 100 institutions in 40 countries.

- **Consumers**: Influencing of consumers to buy products from developing countries can play an important part in addressing the challenges of poverty in these countries, for example, Fair Trade.
- **Investment funds**: There are many socially responsible investment funds on the market with assets in the order of $4 trillion. This is still a small percentage, about 3 per cent, of total global markets – but it is increasing.
- **Business**: The rapid emergence of the new green economy has meant that business is searching for profitable 'green' sustainable opportunities. For example, more than 60 companies are bidding for the rights to secure tidal futures in the Pentlands Firth, Scotland. This would be the world's largest tidal energy project. Similar investments are being made in solar, wind and bio-fuels.
- **International aid**: Provisions made by national governments and large philanthropic institutions is an important source of funding for projects linked to the Global Challenges.
- **Taxation**: Increased taxes on pollution means more money for seeding research and development of green energy. City traffic congestion charges can finance modern efficient transport systems. Government has a major role to play. In Sweden, for example, the government used a whole range of initiatives such as incentives for bio-fuel production, research into energy efficiency and affordable train services as part of their commitment to reduce the use of oil in road transport by 40 per cent by 2020.
- **Remittances**: Funds sent back to developing countries from their overseas citizens are in the order of $300 billion per annum, which is a greater amount than both official aid and private capital investment combined.

- **Sustainable development financial mechanisms.**
    - **Cap and Trade**: Cap and Trade provides an important opportunity to effectively price carbon emissions, and use this financial mechanism to provide funds for innovation in the energy sector.
    - **Micro-banking**: Micro-finance, as demonstrated by the Nobel Prize winning initiatives in Bangladesh, provide an important model for addressing both the poverty trap and developing entrepreneurship. Small loans, given primarily to women, provided an important opportunity for families who would otherwise have no access to funds. This source of funding is now being used in many developing countries.
    - **Wetlands banking**: Other examples of sustainability funds are exemplified by the emergence of the Wetlands Funds in the US, which offset damage to existing wetlands with restored wetlands.[2]
    - **Species banking**: Similar to wetland banking, the US has proved an important area for the development of species banking, where damage to any species can be offset by creating habitats for the same species elsewhere. The benefit is that species and environments that previously had no 'economic value' under the previous models now have a financial value – making consideration of conservation and preservation issues an economic necessity, not an option.

---

[2] As reported in water.usgs.gov/nwsum/WSP2425/legislation.html and State of the World, p. 128.

- **Refocus defence spending**: As already identified, a reallocation of less than 20 per cent of the 2008 defence budget would make a significant contribution to the gap in costs required to address the Global Challenges.[3]
- **Global Fair Trade**: A further key consideration is the establishment of an equitable global trading system. Much work has been done on this, culminating most recently in the DOHA Conference. Though agreement was not reached, all countries acknowledged that such a system must be developed.
- **From 'Greed is Good' to 'Green is Good'**: Most of the national economies on Planet Earth are based on the classic growth model. However, on a planet with finite resources, unending growth is clearly non-sustainable and this is already demonstrated in key areas such shortages of energy, water and food.

Therefore the challenge moving into the future is to progress from a constant growth model to a sustainable economic model – the 'steady state economy' in which economic activities fit into the limited capacity of ecosystems – along with a fairer balance in wealth distribution and the allocation of scarce resources. The heart of the matter is sustainability. The UN Bruntland Commission Report in its 1987 Report 'Our Common Future' defined 'sustainable development' as: 'Development that meets the needs of the present, without compromising the ability of future generations to meet their own needs.' This must be our objective.

The Role of Government, Business and People

GOVERNMENT

National and international government can play important roles in the effective focus, coordination and deployment of funds to address the sustainability requirements of Planet Earth.

INTERNATIONAL GOVERNMENT

One of the first opportunities at an international level is to organise and coordinate resources and funding around the Global Challenges and global fund requirements. This will include key funding organisations including The World Bank, IMF, UN Population Fund, UN Environmental Programme, European Investment Bank, national government and NGOs.

The second and equally important opportunity is to ensure the effective coordination between these groups to ensure the interconnectivity of the Global Challenges are effectively managed. This is one area where significant improvement can certainly be achieved.

Global financial stability: The final area of opportunity for developed countries is the establishment of a stable and predictable economic environment for sustainability. This includes clear guidelines for sustainable investment and direction.

Following the shock of the 2008/9 global financial crisis, it also involves the development and implementation of global financial safeguards to avoid this crisis happening in future.

---

[3] As reported in Brown, L. *Plan B 3.0*, p. 284. Earth Policy Institute, W. W. Norton & Company. 500 Fifth Avenue, N.Y. 10110.

Addressing this can also be integrated with the increased understanding of the meaning of sustainable growth.

With Planet Earth's finite resources, and unlimited growth being clearly unsustainable, an important opportunity lies in laying the foundations for the transition from traditional models of growth economics to sustainable economics.

NATIONAL GOVERNMENT

- **Focus on Top 20**: In setting strategies for financing sustainability we can concentrate on financing sustainability for the top 20 major economic powers on Planet Earth. Funding to address the Global Challenges from these countries is crucial, as well as providing leadership and coordination of global efforts. There is also a key role to play in building a fairer global trade environment that can allow developing countries to leverage their capabilities.

    The G20 countries showed significant leadership and a cooperative approach in addressing the global financial crisis during the 2009 London G20 summit. Now they have the opportunity to demonstrate this same level of leadership and cooperation in addressing all of the Global Challenges and their interconnectivity.
- **Focus on developing countries**: Developing countries have the opportunity to develop and implement plans to address their unique needs. The increased funding allocations to the IMF to support these countries demonstrates recognition of the joint responsibility to address the Global Challenges and support developing countries in achieving this.
- **Defence spending**: The opportunity already outlined is for all countries to commit themselves to a reallocation of at least 20 per cent of defence spending and resources towards 'defending' against the Global Challenges. A joint agreement to this over the next 20 years would provide a ready source of capital and available resources to fast track the actions required. Equally, an agreement to share the allocation across all countries would also minimise the military exposure of any country as each country would be refocusing expenditure on the same basis. Building the cooperation required to achieve this would also build an important platform for joint efforts to address the Global Challenges.
- **Funding allocation**: Funds must be allocated to appropriate Global Challenge budgets. Countries facing food or water shortages can potentially allocate funds to these problems. Countries with high fertility rates can allocate funds to population stabilisation programmes. Countries with surplus funds, using an approach similar to Cap and Trade, could support countries with budgets that do not meet their full requirements.

BUSINESS

The second agent of change, Business, clearly plays a vital role in the financing of sustainability – particularly in the area of economic investment and development.

- **Business planning for the new green revolution**: There is no denying that the changes occurring on Planet Earth in terms of energy costs, global warming, food and water supplies and costs, and population growth are creating an economic and global environment which needs to be properly addressed if any business is to succeed long term.

The role of corporate social responsibility has moved from where it was 20 years ago, a 'nice to have' strategy, to being a key element of any long-term business strategy and necessary for its sustainability. Increasingly companies are appointing CSOs, as these sustainability drivers can no longer be ignored, and in fact need to be proactively addressed by business.

- **Investment in sustainability innovation**: Another key important contribution of business is innovation. Successful innovation will be the key to developing and delivering the technology required to support and drive the new green revolution. Already, much innovation is taking place within Silicon Valley and other key centres of development where there is already a strong focus on new technologies for this revolution.
- **Global financial management mechanisms**: The financial industry plays an essential role here in the development and deployment of rigorous financial management mechanisms to ensure that the funds allocated are spent correctly and effectively.
- **The development of Offset or Cap and Trade mechanism**: This provides a more robust mechanism than has been previously used to provide funding for projects that address the Global Challenges.
- **Defence innovation and deployment**: Business has an important part to play in development and deployment of sustainable defence technology and innovations – particularly in adaptation to Climate Change, water management, food production and resource depletion.

PEOPLE

Voters and shareholders have their part to play in the financing of sustainability.

- **Electoral focus**: In democratic counties, which characterise the G8 Countries, the electorate can use their voting power to elect governments which are committed to financing these challenges. In various ways, such as lobby groups, they can put pressure on politicians to ensure that a long-term sustainability strategy is established and adequately funded.
- **Consumer focus**: As consumers, people have the opportunity to invest their funds in sustainable products and services and to buy from companies that support strong sustainability strategies.
- **Investment in sustainability innovation**: Large business is not the only source of innovation. Companies such as HP, Microsoft and Apple all began as small operations with good ideas – often in a 'garage'. Innovation needs to be encouraged not only for large businesses but also at an individual level and at a small business level.

How Much Will it Cost to Address These Strategies?

Estimating the total cost of financing the work to address the Global Challenges is extraordinarily complex. There are multiple variables, each of which rests on assumptions which are not always universally accepted. However, to commit to objectives and actions without at least working out the scale of the costs is not good management and can be counterproductive.

We are privileged to use the studies made by the world's outstanding leaders in the fields of sustainability. Lester Brown, President of the Earth Policy Institute, has 24 Honours degrees and numerous awards including the UN Environment Prize. Jeffrey Sachs is Director of the Earth Institute and Quetelet Professor of Sustainable Development at Columbia University. He is Special Advisor to the UN Secretary Ban Ki-moon.

Lester Brown, in his book *Plan B 3.0* (Earth Institute, WW Norton & Co, New York, 2008), provides the following analysis:

Goal	Funding ($ billion)
Basic Social Goals	
Universal primary education	10
Eradication of adult illiteracy	4
School lunch programs for 44 poorest countries	6
Assistance to preschool children and pregnant women in 44 poorest countries	4
Reproductive health and family planning	17
Universal basic healthcare	33
Closing the condom gap	3
Total	77
Earth Restoration Goals	
Planting tress to reduce flooding and conserve soil	6
Planting trees to sequester carbon	20
Protecting topsoil on cropland	24
Restoring rangelands	9
Restoring fisheries	13
Protecting biological diversity	31
Stabilizing water tables	10
Total	113
Grand Total	190

Figure 55 Estimated additional costs to address the basic social and earth restoration goals

Jeffrey Sachs in his book *Common Wealth: Economics for a Crowded Planet* (Allen Lane Penguin Group, The Strand, London, 2008) sets out the illustrative annual financial outlays to meet the Millennium Promises.

Global Goal	Financial Need	Illustrative Annual Outlays for Global Cooperation
Climate change mitigation	Adoption of sustainable energy systems, with support for the poorest countries	1.0 per cent of GNP (donor countries) 0.5 per cent of GNP (low-income countries)
Climate change adaptation	Assistance to support the poorest countries with adaptation	0.2 per cent of GNP (donor countries)
Biodiversity conservation	Financing of protected areas	0.1 per cent of GNP (donor countries)
Combating desertification	Financial assistance for water management in low-income dry lands	0.1 per cent of GNP (donor countries)
Stabilizing global population	Assistance for universal access to reproductive health services	0.1 per cent of GNP (donor countries)
Science for sustainable development	Global public financing of research and development of new technologies for sustainable development	0.2 per cent of GNP (donor countries)
Millennium Development Goals	Assistance to help the poorest countries to escape from the poverty trap	0.7 per cent of GNP (donor countries)
Total	Budgetary outlays for global sustainable development	2.4 per cent of GNP (donor countries)

Figure 56   Illustrative annual outlays to achieve the global goals

Putting the Finances of Sustainability in Perspective

These analyses show that the scale of investments is high and will certainly call for greater commitment by the developed countries. Sustained political will and commitment will be required. Why should people make such sacrifices? Simply because the alternative is irreversible damage to Planet Earth. The major impact will be on our grandchildren and their children, with those in the developing world suffering most. The harsh reality is that if Climate Change is not successfully managed, major crop failures, flooding and famine will face the world. If unsustainable population growth is not addressed we move into a world of food and water shortages and conflict over scarce sources.

Yet, there are two reasons to be optimistic as these issues become better understood.

Firstly, historically, when a country or group of countries has been threatened from outside, it forces a genuine sense of common purpose and sacrifice to be created. Survival is a great motivator. These ten challenges threaten us all. We must overcome differences and work together to address them.

Secondly, there is often too much emphasis on costs of sustainability and not enough on the benefits of building a sustainable world. However, we are at the point of an exciting green revolution – a time when we can recreate a world that benefits not only our generation, but all future generations. This is well captured by the words of Jeremy

Oppenheim, Director of Climate Change Initiatives at McKinsey & Co., when speaking at the DAVOS World Economic Forum in 2008.

'The transition to a sustainable economy, if implemented correctly, has the potential to stimulate economic growth, create jobs and bring benefits to consumers. The costs of this transition are in fact investments in new, 21st century infrastructure that will pay off for generations to come – just as the 'costs' of investments in infrastructure such as electrification, highways, and the internet paid off with very high returns for the societies that made them in the 20th century.'

UNDERSTANDING THE 'TRAGEDY OF THE COMMONS'

In 1968, the publication in Science (162:1243–1248, 1986) of the article entitled 'The Tragedy of the Commons', by Gerret Hardin, created expectations about the ability of the human race to manage common areas and resources. Hardin asserted that these common resources cannot be managed without private ownership based on the assumption that people would always maximise their own personal advantage to the disadvantage of the common good.

If we look at areas such as Climate Change, air pollution and deforestation these assumptions appear to be borne out. The human race as a whole is being disadvantaged by the actions of individuals driven by personal advantage.

However, this is not always the case – even when we look at the commons in Medieval Britain. There were many examples of the commons being cooperatively managed so all the commoners benefited. People realised that if the common area was destroyed or was no longer usable then everyone would lose.

Similar examples exist in other cultures. In Bali, for instance, traditional methods of water management had provided for effective common use of the common water resources over centuries. However, in the 1960s 'modern' water management systems were introduced as well as skilled technicians and pesticides. However, crop yields actually decreased with these modern methods. In the end, the traditional methods were reintroduced and crop yields were restored to the previous levels.

This is also borne out by the more recent understanding of game theory, which demonstrates that a gaming strategy of pure self-interest is less successful for human beings than one in which a win-win strategy is used.

The challenge for the human race is which approach will be followed as global resource usage becomes more and more critical? A cooperative win-win approach to the management of Planet Earth – or one driven by self-interest and the maximisation of personal gain – which will inevitably lead to increased conflict in which everybody – including our future generations – lose.

The challenge is for the human race to avoid what in future might be called, not the Tragedy of the Commons, but the Tragedy of Planet Earth.

CHAPTER 12
Executive Brief No. 11: The Challenge of Interconnectivy – The Perfect Storm or the Perfect Opportunity?

'Planet Earth is a spaceship; a beautiful one which took millions of years to develop. It doesn't come with an instruction book. For humanity to survive lastingly and successfully we must see the Planet as a total system and steward its resources accordingly.'

Buckminster Fuller

The Problem of Interconnectivity

In reviewing the ten Global Challenges it becomes increasingly clear that they are closely interconnected and influence one other. For example, global warming influenced by human beings, in turn influences the natural balance of Planet Earth.

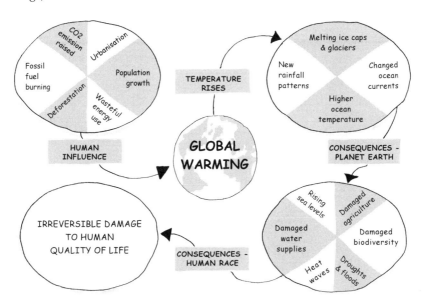

Figure 57   The interconnectivity of global warming and other Global Challenges

Table 8 highlights the cross impact or interconnectivity of some of these ten Global Challenges.

Table 8   The challenges of interconnectivity

Global Challenges	Impact	Some Cross Impacts
Population Growth	Population increasing, particularly in developing countries.	• Increased demands on resources • Increased CO_2 emissions • Increased poverty • Increased conflict over resources
Climate Change	Increased temperature due to greenhouse effect	• Impacts water and food availability • Impacts health, poverty, biodiversity
Energy Supplies	Increasing demand on existing supplies	• Increased food and water costs • Potential increased greenhouse gases if transition to sustainable sources is not fast enough
Water and Food Supplies	Ability of existing water and food supplies to support population increases	• Decreased health • Increased extreme poverty • Increased conflict over resources
Planet Sustainability and Biodiversity	Decreased biodiversity	• Increased greenhouse gas absorption • Decreased food production
Extreme Poverty	Human suffering and economic costs	• Decreased health • Increased demand on resources
Global Health	Limited access to health services and education	• Increased spread of infectious diseases such as HIV/AIDS and Malaria
Universal Education	Limited education reinforces the poverty cycle	• Increased population • Increased poverty • Inefficient use of resources
Conflict and Peace	Conflicts decrease resources and focus on addressing the Global Challenges	• Increased poverty, migration and refugees, water and food supplies • Decreased health and education
Financing a Sustainable World	Inability to finance sustainable strategies impacts all challenges	• Population; Climate Change; Energy, Water and Food Supplies; Sustainability; Poverty; Health; Education; Conflict

The interconnectivities become increasingly self-evident. Success in alleviating extreme poverty creates more people with higher incomes and will result in increased greenhouse gas emissions. The growth of an affluent middle class in China illustrates this. China is now the second largest emitter of greenhouse gases.

The cross impact of the Global Challenges in health is clearly apparent when we look again at the WHO chart. Preventing the impact of major diseases will certainly improve health, particularly in the developing nations. However, these are countries where population growth is already forecast to grow most rapidly, thereby increasing potential poverty and population growth and resource usage.

The WHO chart also illustrates many other health-related issues. Climate Change will increase the spread of diseases. Loss of safe and reliable water reduces the provision of drinking water and the ability to produce food which then increases malnutrition and susceptibility to disease, particularly in children.

The Challenege of Interconnectivity 151

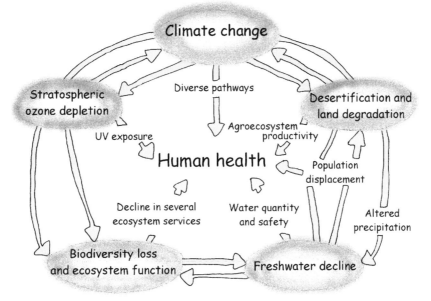

Figure 58 The interconnectivity of human health and other Global Challenges

Greenhouse gas emissions not only influence Climate Change but they are also air pollutants which directly damage health. It is obvious that unless the Global Challenges are managed in a planned, integrated and coordinated manner, fixing one issue may only make another worse.

Biodiversity is influenced by a number of the challenges including Climate Change, pollution and destruction of habitats. These in turn impact other challenges and human well-being. This is captured well in the following UN Environmental Programme table.

Pressures	Impacts on biodiversity	Potential implcations for ecosystem services and human well-being	Examples
Habitat conversion	• Decrease in natural habitat • Homogenization of species composition • Fragmentation of landscapes • Soil degradation	• Increased agricultural production • Loss of water regulation potential • Reliance on fewer species • Decreased coastal protection • Loss of traditional knowledge	Between 1990 and 1997 about 6 million hectares of tropical humid forest were lost annually. Deforestation trends differ from region to region, with the highest rates in Southeast Asia, followed by Africa and Latin America. Additionally, about 2 million ha of forest are visibly degraded each year (Achard and others, 2002).
Invasive alien species	• Competition with and predation on native species • Changes in ecosystem function • Extinction • Homogenization • Genetic contamination	• Loss of traditionally available resources • Loss of potentially useful species • Losses in food production • Increased costs for agriculture, forestry, fisheries, water management and human health	The comb jelly, Mnemiopsis leidyi, accidentally introduced in 1982 by ships from the US Atlantic coasts, dominated the entire marine ecosystem in the Black Sea, directly competing with native fish for food, and resulting in the destruction of 26 commercial fisheries by 1992 (Shiganova and Vodim, 2002).

Figure 59 The impacts of the Global Challenges on biodiversity

Pressures	Impacts on biodiversity	Potential implcations for ecosystem services and human well-being	Examples
Overexploitation	• Extinctions and decreased populations • Alien species introduced after resource depletion	• Decreased availability of resources • Decreased income earning potential • Increased environmental risk (decreased resilience) • Spread of diseases from animals to people	An estimated 1–3.4 million tonnes of wild meat (bushmeat) are harvested annually from the Congo Basin. This is believed to be six times the sustainable rate. The wild meat trade is a large, often invisible contributor to the national economies dependant on this resource. It was recently estimated that the value of the trade in Cote d'Ivoire was US$150 million/year, representing 1.4 per cent of the GNP (Post, 2005).
Climate change	• Extinctions • Expansion or contraction of species ranges • Changes in species compositions and interactions	• Changes in resource availability • Spread of diseases to new ranges • Changes in characteristics of protected areas • Changes in resilience of ecosystems	Polar marine ecosystems are very sensitive to climate change, because a small increase in temperature changes the thickness and amount of sea ice on which many species depend. The livelihoods of indigenous populations living in sub-arctic environments and subsisting on marine animals are threatened, since the exploitation of marine resources is directly linked to the seasonality of sea ice (Smetacek and Nicol, 2005).
Pollution	• Higher mortality rates • Nutrient loading • Acidification	• Decreased resilience of service • Decrease in productivity of service • Loss of coastal protection, with the degradation of reefs and mangroves • Eutrophication, anoxic waterbodies leading to loss of fisheries	Over 90 per cent of land in the EU25 countries in Europe is affected by nitrogen pollution greater than the calculated critical loads. This triggers eutrophication, and the associated increases in algal blooms and impacts on biodiversity, fisheries and aquaculture (De Jonge and others, 2002).

Figure 59   The impacts of the Global Challenges on biodiversity *concluded*

Mexico City, as already highlighted, is an example of what can happen when this type of interconnectivity goes wrong. Its population increased sixfold in the second half of the twentieth century, inevitably creating new demands on the water supplies and land conversion. Water consumption greatly increased and the demand from the aquifers is now greater than the replenishment rate. This is already causing serious water shortages yet the population continues to grow.

To make matters worse, the vast urban sprawl with its roads and buildings created a concrete barrier so that rain can no longer flow through the earth to the aquifers but is diverted to sewage drains. Poor construction means many water pipes leak – 40 per cent of drinking water is lost. The current position is desperate and getting worse, with aquifers at dangerously low levels. Water rationing has been introduced and inevitably water shortage will in turn create a new range of health problems.

Another example is how global Climate Change influences regional water availability, which in turn creates a water crisis which leads to destabilisation, conflict and violence. This is well captured in the chart opposite.

The role of women is an important example of the positive benefits of interconnectivity. There is increasing global action to improve gender equality, fair treatment and empowerment. An important part of this action is to increase educational opportunities as two-thirds of women remain illiterate. However, better education for women does not

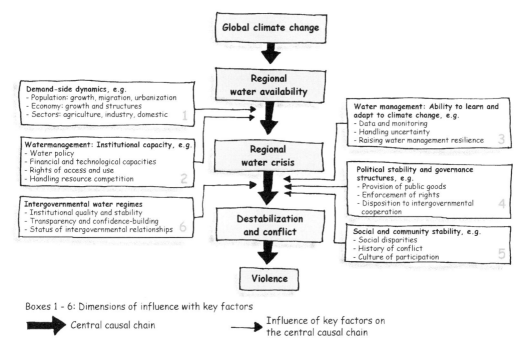

Figure 60 Impacts of climate change on water availability and potential conflict
Source: WBGU.

necessarily lead to greater empowerment when the ownership of wealth is often male and where, in many cultures, women remain the main source of agricultural labour. The link between conflict and violence is becoming more significant in war zones such as the Congo, where rape has become an explicit instrument of war. There are many examples, however, the important lesson is that a holistic approach is the only way to truly give respect to and empower women. Experience has shown that isolated initiatives, no matter how well intentioned, will not work – and could cause more problems than they solve.

Finally, it is essential that interconnectivity is established between all the organisations which are involved in meeting these challenges – at a local, national and global level. There are many examples of multiple governments and NGOs providing very similar aid support to a developing country with little or no coordination. This causes confusion and is clearly a highly inefficient use of resources. The opportunity is, however, for these groups to work together to develop common visions, objectives and priorities and action plans – to leverage each other's resources, experience and skill sets, and demonstrating the positive opportunities of interconnectivity.

The Perfect Storm?

As we have seen, each of the ten Global Challenges on their own pose a significant threat to the human race and the planet. However, combined, they could become the 'perfect storm' threatening the very existence of human civilisation. The key factor is time. If there were another 100 years to address these challenges, then perhaps they could be managed individually. However, many of them need to be managed within the next 10 to 20 years if they are to be dealt with successfully.

Certainly this is true for Climate Change, oil, water and food shortages, as well as sustainability. If the overlay of unrestrained population growth is added then the stability of the human race's financial and economic systems could be further threatened, as well as the existing conflict resolution mechanisms.

Major conflicts and multiple civil wars have a major impact on health of human beings. A population debilitated by World War One was vulnerable to the influenza pandemic which killed up to 100 million people – more people than died in the war itself.[1]

It is very easy to see how quickly these Global Challenges can combine to bring a major breakdown in the quality of life for all on Planet Earth if not quickly and effectively addressed.

The Perfect Opportunity?

The profound sense of urgency and recognition of the need for coordinated global efforts to tackle these Global Challenges presents a major opportunity to work together on an unprecedented scale. The most effective way to galvanise everyone into action is the full recognition of the common threat which the Global Challenges present – a global wake up call!

The first piece of good news is that we are seeing this 'global wake up call' being increasingly recognised, as common challenges such as Climate Change, sustainable energy, food, water and biodiversity – as well as their interconnectivities – become more apparent. Also, the human race has already developed and deployed many of the solutions to address each of the challenges and many examples of 'best practice' are available to benefit from their interconnectivity. And, where solutions are not fully understood, the human race already has the innovation and adaptation capabilities to develop them into practical form.

Table 9   Solutions and cross benefits

Key Challenge	Solution	Impact	Cross Benefit
Population Growth	Female Education Programmes	Educated women have fewer children	Better female health
Climate Change	Reduced CO_2 emissions	Slows global warming	Reduces energy consumption
Energy Supplies	Alternative green energy supplies	Reduces demand on existing supplies	Reduces CO_2 emissions and global warming
Water and Food Supplies	Increased efficiency of food and water production	Better management of existing water and food supplies	Increased health
Planet Sustainability and Biodiversity	Reduced deforestation	Increases biodiversity	Improved greenhouse gas management
Extreme Poverty	Basic education for all	Breaks the poverty cycle	Improved disease prevention
Global Health	Health clinics	Improved access to health services and health education	Reduced spread of infectious diseases such as HIV
Universal Education	Improved literacy programmes	Increased education possible from health to land and water usage	Improved population management Improved sustainability
Financing a Sustainable World	Refocus of defence resources to Global Challenges	Improved health, economic well-being	Impacts ability to address all challenges.
Conflict and Peace Management	Increased mediation	Less escalation to major conflicts	Decreased negative economic and health impacts

---

1   As reported in *Aspirin: The Story of a Wonder Drug*, Jeffreys, D., p. 124, Bloomsbury Publishing Plc, 36 Soho Square, London, W1D 3QY, England.

The overall challenge has been understanding how to tackle these challenges and deploy these solutions and best practices in a coordinated manner, and on a global scale.

The Key Next Steps: Developing a Plan for the Planet

So how can the human race tackle these challenges in a more effective coordinated manner? After all, the current technical and knowledge base that is available means that each of the Global Challenges is understood sufficiently to begin taking rapid action. And where solutions are not fully understood, the human race already has the innovation and adaptation capabilities to develop them into practical form.

The clear challenges moving forward are to get sufficiently focused, organised and disciplined to roll out these solutions globally.

Figure 61   I've got the solution! No, I've got the solution…

The important starting point is the creation of an overall plan to bring together these solutions into a coordinated approach. Fortunately, this is not as difficult as it may at first appear. Many businesses have been successfully developing and implementing complex plans on a global scale. These are based on proven principles – establishing an overall vision for the future and then setting objectives and strategies to address the key challenges and opportunities.

156 Developing a Plan for the Planet

Figure 62   We've all got the solution!

Global business planning principles therefore provide an important framework which can be applied to the Global Challenges being faced on Planet Earth. This will be the focus of the following chapter – the development of a Plan for the Planet.

Since a plan can not be implemented without effective management it is equally important to also look at the global management best practices and how these can be applied to a Plan for the Planet. These are summarised in Chapter 4, together with a series of action checklists that organisations, communities and individuals can use to ensure the effective management of the Global Challenges and their solutions.

Using this approach, we all have the opportunity to turn the potential Perfect Storm into the Perfect Opportunity for the human race and through rapid and coordinated planning and action build a sustainable world for our generation, and all future generations.

PART III
Developing a Plan for the Planet

CHAPTER 13

Building a Global Vision for Planet Earth

'Vision without action is merely a dream. Action without Vision just passes the time. Vision with action can change the world.'

Joel Barker

'Most people do not plan to fail – they fail to plan.'

Anon

Why a Vision for Planet Earth?

Vision setting is often misunderstood and sometimes seen as a 'motherhood statement' or simply a waste of time. However, having a clear long-term Vision is essential in developing any effective plan. There are a number of important reasons for this:

- **A Vision provides an understanding of direction**: When planning for effective change, and particularly rapid change, everyone needs to be a contributor and to be part of the change. And not only to be part of it but to be enthused and passionate about it. To believe in what is happening. This is what truly drives transformational change.

    With change comes opportunity and at no time in the history of the human race has global change been so compelling and significant. A Vision also provides a framework within which objectives and strategies can be developed and implemented. A context within which decisions on implementation can be tested.
- **A Vision provides a long-term context for action**: The second reason to establish a clear Vision is illustrated by the example of the Marshall Plan following the Second World War. The success of the Marshall Plan demonstrates how many countries were mobilised to develop a global rebuilding programme following the destruction and devastation of the war.

    The architects of the Marshall Plan applied many of the elements of global best practice management principles to rebuild Europe following the destruction of economic life. However, the Marshall Plan had a wider Vision: to ensure that the great European powers of Europe did not fight each other again. The foundation was laid for economic unity and this led to the EU. It is now unthinkable that there could be war between the European powers. A Vision realised.
- **A Vision provides an understanding of the world we are operating in**: Most people familiar with global business will know the business term 'marketing myopia'. Blindness to truly understanding 'What business are we really in?' The US Railway industry at the beginning of the 1900s was very successful and seen as a blue chip investment. However, failure to have a Vision that the railways were actually in the transport business, not just the railway business, lead to those companies declining.

- **A Vision provides a basis for a set of Values**: An overall Vision is important, since it sets the framework for the Values that guide decisions. Having a set of Values is crucial. Differing Values create misunderstanding which is not a recipe for effective action.
- **A Vision can provide a basis for survival**: Some of Planet Earth's faith groups put the matter more bluntly, and say a Vision is necessary for survival – 'without a Vision, the people perish'. Perhaps this has never been more relevant.

Building a Vision for Our Plan for the Planet

A Vision for Planet Earth must be based on fundamental enduring principles. The best statement of these principles is contained in the Earth Charter developed over the last decade and presented at the Technical Museum, Stockholm, Sweden, 11 April, 2007 Hard Rain Seminar Series by Dominic Stucker.

Vision Statement for our Plan for the Planet

A planet where there is: respect for nature, observance of universal human rights, economic justice, and a culture of peace.

THE PRINCIPLES ON WHICH THIS VISION IS BASED

To better understand the principles behind our Vision for the Plan for the Planet, a full summary of The Earth Charter has been provided.

THE EARTH CHARTER

Preamble

We stand at a critical moment in Earth's history, a time when humanity must choose its future. As the world becomes increasingly interdependent and fragile, the future at once holds great peril and great promise. To move forward we must recognise that in the midst of a magnificent diversity of cultures and life forms we are one human family and one Earth community with a common destiny. We must join together to bring forth a sustainable global society founded on respect for nature, universal human rights, economic justice, and a culture of peace. Towards this end, it is imperative that we, the peoples of Earth, declare our responsibility to one another, to the greater community of life, and to future generations.

Earth, Our Home

Humanity is part of a vast evolving universe. Earth, our home, is alive with a unique community of life. The forces of nature make existence a demanding and uncertain adventure, but Earth has provided the conditions essential to life's evolution. The resilience of the community of life and the well-being of humanity depend upon preserving a healthy biosphere with all its ecological systems, a rich variety of plants and animals, fertile soils, pure waters, and clean air. The global environment with its finite resources is a common concern of all peoples. The protection of Earth's vitality, diversity, and beauty is a sacred trust.

The Global Situation

The dominant patterns of production and consumption are causing environmental devastation, the depletion of resources, and a massive extinction of species. Communities are being undermined. The benefits of development are not shared equitably and the gap between rich and poor is widening. Injustice, poverty, ignorance, and violent conflict are widespread and the cause of great suffering. An unprecedented rise in human population has overburdened ecological and social systems. The foundations of global security are threatened. These trends are perilous – but not inevitable.

The Challenges Ahead

The choice is ours: form a global partnership to care for Earth and one another or risk the destruction of ourselves and the diversity of life. Fundamental changes are needed in our Values, institutions, and ways of living. We must realize that when basic needs have been met, human development is primarily about being more, not having more. We have the knowledge and technology to provide for all and to reduce our impacts on the environment. The emergence of a global civil society is creating new opportunities to build a democratic and humane world. Our environmental, economic, political, social, and spiritual challenges are interconnected, and together we can forge inclusive solutions.

Universal Responsibility

To realize these aspirations, we must decide to live with a sense of universal responsibility, identifying ourselves with the whole Earth community as well as our local communities. We are at once citizens of different nations and of one world in which the local and global are linked. Everyone shares responsibility for the present and future well-being of the human family and the larger living world. The spirit of human solidarity and kinship with all life is strengthened when we live with reverence for the mystery of being, gratitude for the gift of life, and humility regarding the human place in nature.

We urgently need a shared Vision of basic Values to provide an ethical foundation for the emerging world community. Therefore, together in hope we affirm the following interdependent principles for a sustainable way of life as a common standard by which the conduct of all individuals, organizations, businesses, governments, and transnational institutions is to be guided and assessed.

Principles

I. Respect and Care for the Community of Life

C1. Respect Earth and life in all its diversity.

a. Recognize that all beings are interdependent and every form of life has value regardless of its worth to human beings.
b. Affirm faith in the inherent dignity of all human beings and in the intellectual, artistic, ethical, and spiritual potential of humanity.

C2. Care for the community of life with understanding, compassion, and love.

a. Accept that with the right to own, manage, and use natural resources comes the duty to prevent environmental harm and to protect the rights of people.
b. Affirm that with increased freedom, knowledge, and power comes increased responsibility to promote the common good.

C3. Build democratic societies that are just, participatory, sustainable, and peaceful.

a. Ensure that communities at all levels guarantee human rights and fundamental freedoms and provide everyone an opportunity to realize his or her full potential.
b. Promote social and economic justice, enabling all to achieve a secure and meaningful livelihood that is ecologically responsible.

C4. Secure Earth's bounty and beauty for present and future generations.

a. Recognise that the freedom of action of each generation is qualified by the needs of future generations.
b. Transmit to future generations Values, traditions, and institutions that support the long-term flourishing of Earth's human and ecological communities.

In order to fulfil these four broad commitments, it is necessary to:

II. Ecological Integrity

C5. Protect and restore the integrity of Earth's ecological systems, with special concern for biological diversity and the natural processes that sustain life.

a. Adopt at all levels sustainable development plans and regulations that make environmental conservation and rehabilitation integral to all development initiatives.
b. Establish and safeguard viable nature and biosphere reserves, including wild lands and marine areas, to protect Earth's life support systems, maintain biodiversity, and preserve our natural heritage.
c. Promote the recovery of endangered species and ecosystems.

d. Control and eradicate non-native or genetically modified organisms harmful to native species and the environment, and prevent introduction of such harmful organisms.
e. Manage the use of renewable resources such as water, soil, forest products, and marine life in ways that do not exceed rates of regeneration and that protect the health of ecosystems.
f. Manage the extraction and use of non-renewable resources such as minerals and fossil fuels in ways that minimize depletion and cause no serious environmental damage.

C6. Prevent harm as the best method of environmental protection and, when knowledge is limited, apply a precautionary approach.

a. Take action to avoid the possibility of serious or irreversible environmental harm even when scientific knowledge is incomplete or inconclusive.
b. Place the burden of proof on those who argue that a proposed activity will not cause significant harm, and make the responsible parties liable for environmental harm.
c. Ensure that decision making addresses the cumulative, long-term, indirect, long distance, and global consequences of human activities.
d. Prevent pollution of any part of the environment and allow no build-up of radioactive, toxic, or other hazardous substances.
e. Avoid military activities damaging to the environment.

C7. Adopt patterns of production, consumption, and reproduction that safeguard Earth's regenerative capacities, human rights, and community well-being.

a. Reduce, reuse, and recycle the materials used in production and consumption systems, and ensure that residual waste can be assimilated by ecological systems.
b. Act with restraint and efficiency when using energy, and rely increasingly on renewable energy sources such as solar and wind.
c. Promote the development, adoption, and equitable transfer of environmentally sound technologies.
d. Internalize the full environmental and social costs of goods and services in the selling price, and enable consumers to identify products that meet the highest social and environmental standards.
e. Ensure universal access to health care that fosters reproductive health and responsible reproduction.
f. Adopt lifestyles that emphasize the quality of life and material sufficiency in a finite world.

C8. Advance the study of ecological sustainability and promote the open exchange and wide application of the knowledge acquired.

a. Support international scientific and technical cooperation on sustainability, with special attention to the needs of developing nations.
b. Recognize and preserve the traditional knowledge and spiritual wisdom in all cultures that contribute to environmental protection and human well-being.
c. Ensure that information of vital importance to human health and environmental protection, including genetic information, remains available in the public domain.

III. Social and Economic Justice

C9. Eradicate poverty as an ethical, social, and environmental imperative.

a. Guarantee the right to potable water, clean air, food security, uncontaminated soil, shelter, and safe sanitation, allocating the national and international resources required.

b. Empower every human being with the education and resources to secure a sustainable livelihood, and provide social security and safety nets for those who are unable to support themselves.

c. Recognize the ignored, protect the vulnerable, serve those who suffer, and enable them to develop their capacities and to pursue their aspirations.

C10. Ensure that economic activities and institutions at all levels promote human development in an equitable and sustainable manner.

a. Promote the equitable distribution of wealth within nations and among nations.

b. Enhance the intellectual, financial, technical, and social resources of developing nations, and relieve them of onerous international debt.

c. Ensure that all trade supports sustainable resource use, environmental protection, and progressive labour standards.

d. Require multinational corporations and international financial organizations to act transparently in the public good, and hold them accountable for the consequences of their activities.

C11. Affirm gender equality and equity as prerequisites to sustainable development and ensure universal access to education, health care, and economic opportunity.

a. Secure the human rights of women and girls and end all violence against them.

b. Promote the active participation of women in all aspects of economic, political, civil, social, and cultural life as full and equal partners, decision makers, leaders, and beneficiaries.

c. Strengthen families and ensure the safety and loving nurture of all family members.

C12. Uphold the right of all, without discrimination, to a natural and social environment supportive of human dignity, bodily health, and spiritual well-being, with special attention to the rights of indigenous peoples and minorities.

a. Eliminate discrimination in all its forms, such as that based on race, colour, sex, sexual orientation, religion, language, and national, ethnic or social origin.

b. Affirm the right of indigenous peoples to their spirituality, knowledge, lands and resources and to their related practice of sustainable livelihoods.

c. Honor and support the young people of our communities, enabling them to fulfil their essential role in creating sustainable societies.

d. Protect and restore outstanding places of cultural and spiritual significance.

IV. Democracy, Nonviolence, and Peace

C13. Strengthen democratic institutions at all levels, and provide transparency and accountability in governance, inclusive participation in decision making, and access to justice.

a. Uphold the right of everyone to receive clear and timely information on environmental matters and all development plans and activities which are likely to affect them or in which they have an interest.

b. Support local, regional and global civil society, and promote the meaningful participation of all interested individuals and organizations in decision making.

c. Protect the rights to freedom of opinion, expression, peaceful assembly, association, and dissent.

d. Institute effective and efficient access to administrative and independent judicial procedures, including remedies and redress for environmental harm and the threat of such harm.

e. Eliminate corruption in all public and private institutions.
   f. Strengthen local communities, enabling them to care for their environments, and assign environmental responsibilities to the levels of government where they can be carried out most effectively.

C14. Integrate into formal education and life-long learning the knowledge, Values, and skills needed for a sustainable way of life.

   a. Provide all, especially children and youth, with educational opportunities that empower them to contribute actively to sustainable development.
   b. Promote the contribution of the arts and humanities as well as the sciences in sustainability education.
   c. Enhance the role of the mass media in raising awareness of ecological and social challenges.
   d. Recognize the importance of moral and spiritual education for sustainable living.

C15. Treat all living beings with respect and consideration.

   a. Prevent cruelty to animals kept in human societies and protect them from suffering.
   b. Protect wild animals from methods of hunting, trapping, and fishing that cause extreme, prolonged, or avoidable suffering.
   c. Avoid or eliminate to the full extent possible the taking or destruction of non-targeted species.

C16. Promote a culture of tolerance, nonviolence, and peace.

   a. Encourage and support mutual understanding, solidarity, and cooperation among all peoples and within and among nations.
   b. Implement comprehensive strategies to prevent violent conflict and use collaborative problem solving to manage and resolve environmental conflicts and other disputes.
   c. Demilitarize national security systems to the level of a non-provocative defence posture, and convert military resources to peaceful purposes, including ecological restoration.
   d. Eliminate nuclear, biological, and toxic weapons and other weapons of mass destruction.
   e. Ensure that the use of orbital and outer space supports environmental protection and peace.
   f. Recognize that peace is the wholeness created by right relationships with oneself, other persons, other cultures, other life, Earth, and the larger whole of which all are a part.

The Way Forward

As never before in history, common destiny beckons us to seek a new beginning. Such renewal is the promise of these Earth Charter principles. To fulfil this promise, we must commit ourselves to adopt and promote the Values and objectives of the Charter.

This requires a change of mind and heart. It requires a new sense of global interdependence and universal responsibility. We must imaginatively develop and apply the Vision of a sustainable way of life locally, nationally, regionally, and globally. Our cultural diversity is a precious heritage and different cultures will find their own distinctive ways to realize the Vision. We must deepen and expand the global dialogue that generated the Earth Charter, for we have much to learn from the ongoing collaborative search for truth and wisdom.

Life often involves tensions between important Values. This can mean difficult choices. However, we must find ways to harmonize diversity with unity, the exercise of freedom with the common good,

> short-term objectives with long-term goals. Every individual, family, organization, and community has a vital role to play. The arts, sciences, religions, educational institutions, media, businesses, nongovernmental organizations, and governments are all called to offer creative leadership. The partnership of government, civil society, and business is essential for effective governance.
>
> In order to build a sustainable global community, the nations of the world must renew their commitment to the United Nations, fulfil their obligations under existing international agreements, and support the implementation of Earth Charter principles with an international legally binding instrument on environment and development.
>
> Let ours be a time remembered for the awakening of a new reverence for life, the firm resolve to achieve sustainability, the quickening of the struggle for justice and peace, and the joyful celebration of life.

Developing a '2030 Mission'

A Vision requires action to be realised, and therefore the time has now come for action – not just investigations and rhetoric. We can never understand all the issues or all the facts but effective managers understand that at critical stages decisions must be made on the basis of the best existing knowledge and practices and action taken. The solutions and actions can then be monitored, refined and adjusted during implementation to achieve continuous improvement. It is an iterative learning process. This is the process we must now employ if we are to address the urgency that the global challenges present to us.

Effective action will happen on the basis of understanding, effective implementation and management and then continuous improvement. This can be translated into a Mission – driven by established timeframes, and clear deliverables – to ensure our Vision is realised. Using these principles we can establish and encourage everyone to sign up to a common 2030 Mission to achieve a sustainable Planet Earth.

> **A 2030 Mission for a Sustainable Planet Earth**
>
> - To achieve global understanding and commitment to a coordinated, global Plan for the Planet to address the key challenges
>
> - To continuously improve, implement and manage our global Plan for the Planet to achieve the long-term vision for a sustainable planet earth

CHAPTER 14
Global Objectives and Strategies: Addressing Our Key Global Challenges

Our Vision establishes our overall ambition and our mission provides a framework for action. However, the failure to translate these into set objectives leads to confusion, a lack of direction and most importantly, excuses for lack of action. Setting clear objectives, strategies and Action Plans – which everyone can take a part in implementing – is therefore the essential next step in establishing an effective Plan for the Planet.

Fortunately, we can draw on the good work of many organisations. The Millennium Development Goals developed by the UN provide an important framework that can be used for setting objectives for many of the key challenges. These goals were adopted by 189 nations and are outlined in Appendix 1. We can also draw on the research and experience of experts in the field of Climate Change and our other challenges to establish the set of objectives and strategies required to successfully address these global challenges.

Objectives, strategies and action plans will require effective involvement and implementation to be successful. This involvement will require everybody to act – and to act in a coordinated way.

With the global challenges there is no longer a choice between bystander and action. We are all part of the cause and we can all be part of the solutions. In building a sustainable planet we all have to become 'agents of change' whether in government, in business or in the community. This is the benefit of an overall Plan for the Planet. It provides a framework for action and for global coordination.

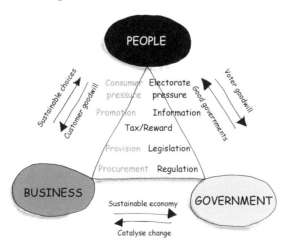

Figure 63   The key agents in the Triangle of Change
Source: Sustainable Development Commission, UK.

168 Developing a Plan for the Planet

The Triangle of Change (shown on the previous page) developed by the UK Department of Environment, Food and Rural Affairs (DEFRA) provides a useful summary of the interactions between each of the key agents of change: Government, Business and People.

This then provides our basis for coordination. To achieve this, our objectives, strategies and Action Plan templates have therefore been developed for each of these key agents of change.

- **Government**: Government at both international and national levels clearly play a key role in establishing a stable and predictable environment in which objectives are pursued through regulatory and policy frameworks.
- **Business**: We are able to use the resources, financial strength and innovation of business and also to learn from and leverage their business and management skills.
- **People**: The third component of the Triangle of Change is people – in their role as contributors to resources, consumers of Planet Earth's resources, investors in business enterprises, electors of governments and as part of the broader communities of which they are a part. Through the electoral process and lobby groups, people influence political policies. Through their purchasing power and investment decisions people influence business.

Figure 64 **No, it's their responsibility!**

Success depends on everyone embracing the objectives, strategies and Action Plans and putting them into practice.

Figure 65 **It's all of our responsibilties**

Involvement of the Global Community

> 'The defining challenge of the 21st Century will be to face the reality that humanity shares a common fate on a crowded planet.'
>
> Jeffrey Sachs

Defining Global Vision and Values calls for cooperation and the opportunity to work closely together. For effective change management to take place the three agents of change – Government, Business and People – need to be committed to – and delivering to a 'common game plan'.

This is the purpose of a Plan for the Planet. To provide a common framework to which everyone – whether in Government, Business or the community – can be involved and contribute. We all become part of the solution.

> 'The survival of our species requires a complete co-operation between all communities of the world. Nations of the world have co-operated together to fight terrorism and they have (and need to further) co-operate to fight against an even greater threat: the destruction of the global life-support systems.'[1]

So many grand Visions are not backed up by detailed Action Plans and effective implementation and thus disappoint and fail. The next section is the therefore the translation of our Plan for the Planet Vision and Mission into our common set of objectives, strategies and Action Plans for each of the key challenges.

Objectives, Strategies and Action Plan Summary

BUSINESS PLANNING: WHAT, WHO, WHEN, AND HOW

Business planning provides a very practical and pragmatic perspective on driving change. To achieve this, our Vision and Mission can be translated into objectives, which can then be turned into strategies, and strategies turned specific actions, with specific responsibilities and timeframes for implementation. An effective globally coordinated plan, translated from Vision to specific Actions Plans is the most effective way of delivering effective results. As we move into the more detailed approach, let's review why each of these elements of good business planning are important.

- **Vision and Mission**: Our Vision and Mission provide the overall direction and reason for action.
- **Objectives, Strategies and Actions**: Our Vision and Mission are then translated into specific Objectives, Strategies and Actions.
- **Ownership**: Actions are established with clear owners – someone taking clear responsibility for each action.
- **Outcomes**: Each action requires clearly defined outcomes or deliverables as a result of that action.

---

1 As reported on http://globalcommunitywebnet.com/globalcommunity/visionearthyear2024.htm.

- **Involvement**: Everybody needs to be involved in driving the change required to implement a successful Plan for the Planet – Government, Business and People.

To translate this into practice, a Business Plan Summary has been provided for each of the global challenges. These have been presented as a series of templates which can then be used as a framework for developing a Plan for the Planet for each of the agents of change. These Objectives, Strategies and Actions have been built taking into consideration the Millennium Development Goals and other key inputs identified in the earlier opportunities and best practices sections, as well as the authors own views. These business-planning summaries are intended to provide a framework from which a Plan for the Planet could be developed, and applied through the keys agents of change – Government, Business and People.

OBJECTIVES AND STRATEGIES

No	Objectives (What are we aiming to achieve?)	Strategies (How are we going to achieve this?)	Actions (What do we have to do to achieve this?)	Outcomes (Results)
1.0	Achieve a sustainable Planet Earth	Develop a Plan for the Planet to provide a framework for each for the agents of to build Action Plans	Each agent of change to develop and implement their Action Plans for achieving sustainability on Planet Earth	Global, coordinated actions through each of the agents of change
		Achieve commitment to implement a global Plan for the Planet	Promotion of the Plan for the Planet to gain understanding and commitment to action	Each of the key agents of change engaged in our Plan for the Planet – Government, Business, People

Business Plan Summary

DEVELOPING A PLAN FOR THE PLANET – OBJECTIVES AND STRATEGIES

No	Objectives (What are we aiming to achieve?)	Strategies (How are we going to achieve this?)	Actions (What do we have to do to achieve this?)	Outcomes (Results)
1.0	Population • A peak population of 8 billion people by 2050 • Identify and achieve a long-term sustainable and equitable population • Achieve by 2015 universal access to reproductive health	International Government • International proactivity in monitoring and providing consultancy and funding to enable national actions to address national population stabilisation strategies	• Monitor growth and support global and national population stabilisation programmes • Source and focus funding to address shortfalls to enable national population stabilisation programmes	Achieve low population forecast growth rates to 2042 Achieve population stabilisation at a sustainable level in the long term.
		National Government • Establish a national family planning programme covering: – education – health clinics – culture – female literacy – religious support	• Monitor growth and support national population stabilisation programmes • Source and focus funding to address shortfalls to enable national population stabilisation programmes	Achieve country-specific targeted population stabilisation objectives
		Business • Communication and Education: contribution to national strategies on family planning • Health Care: Incorporation of family planning into health care programmes	• Create and support local and international initiatives addressing family planning and health clinics	Increased levels of required family planning enabled in appropriate countries
		People • Community Support programmes • Health Care and Social Security programmes	• Take actions to support family planning, community support, female education, health care and social security programmes in developing countries with high forecast population growth	Increased levels of required family planning enabled in appropriate countries

No	Objectives (What are we aiming to achieve?)	Strategies (How are we going to achieve this?)	Actions (What do we have to do to achieve this?)	Outcomes (Results)
2.0	Climate Change • Limit temperature increase to 2°C • Greenhouse gas reduction of 80% by 2050 • Return to preindustrial levels of greenhouse gases long term	International Government • Focused programme targeting the top CO_2 emitting countries • Establish international coordination on Climate Change objectives and initiatives based on the Bali and Copenhagen Agreements and post Copenhagen Agreements for all countries	• Agree targets for greenhouse gas reductions • Establish and agree international and national initiatives to achieve these objectives • Agree long-term initiatives to return to preindustrial levels	e.g. 20% reduction by 2020, 80% reduction by 2050 Return to pre-industrial levels of CO_2 long term
		National Government • National Government creates the right environment for the required change	• Introduction of Cap and Trade system based to achieve greenhouse gas reduction objectives • Introduce "eat your own dog food" programmes within government to reduce government footprint • Reforestation programmes	Monitor against 80% reduction by 2050 targets
		Business • Reduce business carbon footprint to meet global and national objectives • Innovation and Finances: Innovate and invest in CO_2 reduction strategies	• Audit existing CO_2 footprint, direct, and with suppliers • Avoid: i.e. teleconference vs air travel • Reduce: Reduce CO_2 footprint consumption i.e. turn off computers overnight • Replace: Replace high CO_2 footprint equipment i.e. lighting • Offset: Offset where other actions not possible	Knowledge and understanding of CO_2 and greenhouse gas impact Reduction of CO_2 impact

No	Objectives (What are we aiming to achieve?)	Strategies (How are we going to achieve this?)	Actions (What do we have to do to achieve this?)	Outcomes (Results)
2.0 (cont)		People • Reduce household Carbon Footprint to achieve and exceed global and national objectives by: – 50% over next 5 years – 80% over 40 years	Developing country example: • Move to sustainable energy supplier for electricity supplies • Offset all air travel • Move to hybrid car or halve engine size over next 5 years • Heater Boiler replacement in next 10 years. • Become involved Plant a Tree Programmes and initiatives	Potential reductions 12% reduction 30% reduction 10% reduction 8% reduction
3.0	Energy Supplies • Transition to long-term sustainable and non carbon-based energy sources • Maintain the 80% reduction objective in greenhouse gas emissions.	International Government • Support sustainable energy production capabilities through sharing skills and knowledge sharing of global best practice	• Monitor sustainable energy production and best practice • Support best practice roll out of sustainable energy supplies on a global scale	Fast-track growth of sustainable energy technologies and applications
		National Government • National government programme driving for increased energy efficiency	• Introduction of stricter (CAFÉ) standards for motor vehicles • Improved best practice appliance labeling standards • Improved global best proactive building standards • Regulation to focus on lower energy usage in manufacturing and agriculture • Introduce aviation fuel taxes • Refocus subsidies to sustainable energy supplies	Measures of increased energy efficiency and sustainability

No	Objectives (What are we aiming to achieve?)	Strategies (How are we going to achieve this?)	Actions (What do we have to do to achieve this?)	Outcomes (Results)
3.0 (cont)		Business • Move to sustainable energy supplier • Reduce overall energy consumption through elimination of energy waste programmes	• Audit existing energy footprint, directly and with suppliers • Avoid: i.e. teleconf vs air travel • Reduce: Reduce energy footprint consumption i.e. turn off computers overnight • Replace: Replace high energy footprint equipment i.e. lighting • Offset: Offset where other actions not possible	Knowledge and understanding of energy usage impact Reduction of energy impact

		People
		• Reduce household energy consumption by:
		   – 50% over next 5 years
		   – 80% over 40 years
			Developed country example:
			• Avoid – Use virtual shopping
			• Reduce – Turn off appliances
			• Replace – Sustainable energy supplier, hybrid or electric vehicle, more efficient boiler, air drying clothes
			• Offset – Car and air travel
			• Watch and Adapt – Ongoing technology improvements
				Potential reductions
				
				Reduce shopping energy consumption
				
				25% less consumption
				20%+ reduction
				
				40%+ reduction

4.0	Water and Food Supplies • Establish long-term water security at national and international levels • Establish long-term sustainable food supplies • To halve the proportion of people who suffer from hunger over the next 10 years	International Government • International cooperative agreements for integrated water management • Efficiency standards and labeling of appliances • International cooperation on food for all and ocean management	• Monitor sustainable water efficiency and food production best practice • Support best practice roll out of sustainable water efficiency and food production on a global scale	Fast-track growth of water efficiency and food production technologies and applications

Global Objectives and Strategies

No	Objectives (What are we aiming to achieve?)	Strategies (How are we going to achieve this?)	Actions (What do we have to do to achieve this?)	Outcomes (Results)
4.0 (cont)		National Government • Regulation of water usage to achieve sustainable levels of usage • Regulation and international cooperation to ensure adequate and sustainable food supplies	• Improved water use in cities, ground water and river resources, drought management, water purification systems and pollutant management. • Improved food production management, i.e. fisheries	Measures of increased water efficiency and food production sustainability
		Business • Implement best practice water efficiency usage • Implement best practice food production methodologies	• Review and identify industry best practice for water efficiency and food production • Implement into existing operations • Monitor and implement continuous improvement programmes	Measurable improvements in water efficiency and food production
		People • Reduce household water usage by 25% by 2020 • Purchase appropriately certified selection of food	• Replace toilet flush systems with duel flush technologies • Replace shower heads with more efficient type • More selective purchase of food i.e. organic, Fair Trade • Introduce organic farming where viable and appropriate	25% total reduction through application of both strategies over next 10 years Increased viability of sustainable food suppliers
5.0	Planet Sustainability • Integrate principles of country sustainable development • Reverse the loss of environmental and biodiversity resources	International Government • Strengthen international cooperation on reduction of global deforestation, protection of endangered species, desert reclamation and reforestation	• Support key deforestation reduction initiatives • Support key endangered species protection initiatives • Drive desert reclamation and reforestation initiatives	Reversal of deforestation trends Species loss trends reversed. Increase in desert reclamation and reforestation

No	Objectives (What are we aiming to achieve?)	Strategies (How are we going to achieve this?)	Actions (What do we have to do to achieve this?)	Outcomes (Results)
5.0 (cont)		National Government • Establish a national sustainability initiative	• Focused programmes to stop deforestation. Sourcing from managed plantations • Reforestation programmes • Wildlife and endangered species protection programmes • Desert reclamation programmes	Monitoring forestation levels, wildlife management and desert reclamation
		Business • Implement sustainable sourcing, manufacture and production best practice • Implement recycling programmes to increase efficiency and reduce waste	• Review and identify industry sustainability best practice • Implement, with continuous improvement, sustainability best practices across business	Accreditation measures of both internal and external suppliers
		People • Consumers – Focus on consumption of sustainable wood, plant and animal products • Investors – Invest in companies with proven sustainability strategies	• Purchase products with clear certification, low carbon footprint and air miles • Ensure clear and auditable sustainability strategies are in place and being actioned • Support endangered species programmes.	Increase demand for sustainable products increases viability Investment demand for CR increases accountability of global corporate companies
6.0	Extreme Poverty • Halve extreme poverty over the decade, and eliminate extreme poverty over the next two decades. • Achieve poverty reduction programmes in key countries	International Government • Continue and strengthen MDG initiatives to ensure goals are achieved on time • Strengthen monitoring and enforcement of financial regulation on aid programmes • Ensure required financial resources are deployed	• Focus on top 40 Countries where extreme poverty is having highest impact • Establish and implement an MDG improvement plan to ensure that MDG's are achieved on schedule • Extension of debt cancellation	Monitor as per MDG measures already in place

Global Objectives and Strategies

No	Objectives (What are we aiming to achieve?)	Strategies (How are we going to achieve this?)	Actions (What do we have to do to achieve this?)	Outcomes (Results)
6.0 (cont)		National Government • National programmes to address extreme poverty in targeted countries	• Fix the 'leaky pipes' programmes in both donor and recipient countries • Country programmes including reproductive health and planning, education programmes and basic health care	MDG Measures of percentage of aid reaching end users Measures of family planning, education and health care access
		Business • Investment in developing countries with focus on social and community infrastructure • Investment in competitive equality • Audit of all investment to ensure 'no leaky pipes'	• Investment in community infrastructure which address the poverty cycle including primary education, basic health care and reproductive health and planning	Measures focused on increasing education, health care and family planning
		People • Consumers: Focus on consumption of Fair Trade wood, plant and animal products • Investors: Invest in charities and aid programmes with clear and accountable audit paths	• Purchase products with clear fair trade certification • Ensure clear audit strategies are in place and being actioned for all aid donations • Support Country Development Programmes in key countries	Increased free trade Donations end up where required Skills and resource transfer
7.0	Global Health • Ensure safe and reliable water supplies • Eradiate major infectious and lifestyle diseases • Reduce mortality rate in key population groups.	International Government • Continue and strengthen MDG initiatives to ensure goals are achieved on time • Strengthen monitoring and enforcement of financial regulation on aid programmes • Ensure required financial resources are deployed	• Focus on top 40 Countries where health issues are having highest impact • Establish and implement an MDG improvement plan to ensure that MDG's are achieved on schedule	Monitor as per MDG measures already in place

No	Objectives (What are we aiming to achieve?)	Strategies (How are we going to achieve this?)	Actions (What do we have to do to achieve this?)	Outcomes (Results)
7.0 (cont)		National Government • National programmes to address universal health in targeted countries	• Establish national water management strategies • Implement programme to address infectious diseases • Address lifestyle issues impacting health • Ensure basic health care services are established	MDG Measures of percentage of water management, infectious diseases, lifestyle disease management and health care access
		Business • Provision of services: Investment in health infrastructure as part of community investment programme • Tobacco-free environment	• Investment in staff welfare and community support programmes • Provision of philanthropic services and supplies in developing countries	Measures focused on increasing health care and family planning in investment communities
		People • Support safe and reliable water programmes • Support the Tobacco-Free Environment Initiative • Implement lifestyle changes to improve general health	• Support charity and aid programmes focused on safe water supplies and health clinics in key countries • Initiate lifestyle changes as required	Increased resources available to support key water and clinic initiatives Increased awareness and wellbeing
8.0	Global Education • Ensure access to primary education for children everywhere by 2015 • Eliminate gender disparity. • Address illiteracy	• Continue and strengthen MDG initiatives to ensure goals are achieved on time • Strengthen monitoring and enforcement of financial regulation on aid programmes • Ensure required financial resources are deployed	• Focus on top 40 Countries where lack of access to education is having highest impact • Establish and implement an MDG improvement plan to ensure that MDG's are achieved on schedule	Monitor as per MDG measures already in place
		National Government • National programmes to address primary education and illiteracy in targeted countries	• Funding primary education • School lunch programmes • Focus on female education • Education linked teaching scholarships.	MDG Measures of percentage of aid reaching end users Measures of primary education

Global Objectives and Strategies

No	Objectives (What are we aiming to achieve?)	Strategies (How are we going to achieve this?)	Actions (What do we have to do to achieve this?)	Outcomes (Results)
8.0 (cont)		Business • Employment in developing countries • Community Education programmes as part of corporate responsibility actions.	• Investment in employee education programmes • Investment in community education programmes focused on increasing primary education in developing countries and literacy levels	Measures focused on increasing education levels in investment communities
		People • Funding and resourcing – Focus on funding education aid programmes which increase both primary education and literacy levels particularly in developing countries	• Supporting programmes aimed at increasing global literacy levels • Supporting programmes that increase access to primary education in developing countries	MDG measures of global literacy and primary education levels
9.0	Global Peace • Strengthen the UN Peacekeeping capabilities • Strive for the elimination of weapons of mass destruction • Minimise global conflicts for 2010 to 2030 transition period	International Government • Invest in Conflict Resolution and the Peace Process • Invest in Peace Building • Invest in Peacekeeping • Refocus defence priorities • Refocus % of defence spending to address the Global Challenges	• Scale up international peacekeeping and conflict management best practice at a global level • Scale up investment in global peacekeeping • Scale up centres of excellence and expertise in conflict management and peacekeeping	Reduce number of global conflicts and costs associated 2011–2031 (20 year) period of peace in which to focus global efforts and resource on addressing the Global Challenges
		National Government • National conflict management programmes as well as defence programmes • Invest in Conflict Resolution and the Peace Process • Invest in Peace Building • Invest in Peacekeeping • Refocus defence priorities • Refocus defence spending	• Scale up national peacekeeping and conflict management best practice at a global level • Scale up investment • Scale up centres of excellence and expertise	Increased levels of conflict resolution

No	Objectives (What are we aiming to achieve?)	Strategies (How are we going to achieve this?)	Actions (What do we have to do to achieve this?)	Outcomes (Results)
9.0 (cont)		Business • Refocus on Global Challenge Defence and Conflict Management Development • Participate where feasible and appropriate in post conflict physical and economic rebuilding programmes. • Address bribery and corruption issues	• Understand and implement Global Compact conflict resolution and prevention efforts	Reduced conflicts in region
		People • Political lobbying: Support programmes aimed at conflict reduction over next 20 years • Better conflict management – Support local and national conflict management	• Join conflict reduction lobby groups • Develop skills and expertise in personal, community and global conflict management and mediation	Reduced global, national, regional and local conflicts
10.0	Financing Sustainability • Establish global funding strategy for the ten Global Challenges • Establish open and fair trading system	International Government • Leverage international financial mechanisms to fast-track the financing of sustainability • Refocus % of global defence spending from 2010–2030 to addressing the Global Challenges.	• Strengthen G8 and G20 initiatives to address global financial challenges and sustainability financing • Establish cooperative agreement to refocus defence spending for 2010–2030 period	Measures of achievement of resolution of global challenges
		National Government • Establish national financial stability to build growth economy • Focus defence spending to include addressing the global challenges	• Introduce measures to finance sustainability • Allocation of 20% of defence spending to address the global challenges	Measures of sustainability finance required vs actual Measure of financial stability and growth Measures of defense spending allocation

No	Objectives (What are we aiming to achieve?)	Strategies (How are we going to achieve this?)	Actions (What do we have to do to achieve this?)	Outcomes (Results)
10.0 (cont)		Business • Implement Cap & Trade mechanisms • Defence innovation focused on defence against the ten Global Challenges.	• Support for Cap & Trade • Investment focus and innovation to address the ten Global Challenges	Measures focused on global financial stability and KPI's against 10 Global Challenges
		People • Electorate: Support political action to endorse defence spending to deal with the global challenges • Investment: Focus investment in companies that address the global challenge	• Vote and lobby for defence spending on global challenges • Invest in companies that are supporting spending on global challenges	Specific measures of country by country defence allocation to global challenges

YES WE CAN.

President Obama, 2008

I have the great hope that we will have the courage to embrace the changes necessary to save our economy, our planet and ultimately ourselves.

Al Gore (Co-recipient Nobel Peace Prize 2007)

These summary Objectives, Strategies and Action Plans demonstrate that in spite of the complexity of the ten global challenges and their interconnectivity there is a firm understanding of the issues and the solutions. We know much of what has to be done, and who has to do it.

It is easy at this stage however, to feel overwhelmed and to say: 'politicians won't have the will to make these changes', 'self interest will always prevent the essential cooperation required to succeed' and so on. If this negative thinking is accepted, then indeed nothing will be achieved.

However, the human race has demonstrated remarkable resilience throughout the changes and challenges that it has faced on Planet Earth. And these challenges have been overcome through cooperation. The new challenges require cooperation on an unprecedented global scale.

The words of US President Obama have been an inspiration to many. 'Yes we can' captures that spirit of global cooperation, of putting aside the partisan interests that threaten to divide.

The stakes are so high, nothing else will do. No Plan for the Planet can be perfect or complete, therefore every leader of every public and private organisation and every individual has a part to play in improving this Plan for the Planet and making it their own.

However, three complementary actions are essential.

Communication Plan: A global, creative and sustained communication effort must be launched so that everyone is fully aware of the potential threats and opportunities. The plain fact is that many people are not interested or involved because they are not aware of the issues or their implications. Or they are preoccupied with a single challenge such as climate change, without seeing the relationship to the other challenges such as population, food, water, resources and conflict.

Rapid Response: Faced with the global financial crisis in 2008/9, the speed of response, the remarkable cooperation between developed and developing countries and the commitment to find ways to prevent the issues returning – is an important example of what can be achieved in global cooperation when the challenges are understood and faced together.

Effective Management: However, success will only be achieved if these commitments are effectively managed to effective implementation. This is often where grand commitments fail. In the case of the global challenges however, the human race has no room for failure.

Ineffective management in many key organisations –international and national governments, business and charities alike – is a major obstacle to addressing the challenges. Wasteful bureaucracy, lack of coordination, preoccupation with inputs and not results are commonplace and need to be addressed quickly due to the aggressive timeframes for action that the global challenges demand.

For example: with the leadership of the UN, multiple task forces of respected management experts could study the management effectiveness of major organisations and make recommendations. Visionary goals and objectives are of little value unless good management converts them to reality. The opportunity is not only to establish a global plan, but also global best practice management to ensure it is effectively implemented and a sustainable world achieved.

A top down, bottom up approach: This is necessary to achieve the required results in the limited timeframes. There can be no more its 'somebody else's' responsibility. Everyone has the opportunity to take responsibility for the areas they can address in the Plan for the Planet.

Tackling the challenges is everybody's responsibility. In addition, key influencers and opinion formers in society, such as politicians, business and community leaders have an important opportunity to fully understand the issues and give leadership to the solutions.

In parallel the immense contribution of individuals, charities, faith groups and the like which already exist must be encouraged. Such organisations may often be small in themselves but in total have an exceptional record of achievement.

TOGETHER WE CAN!

President Obama

CHAPTER 15
Taking Responsibility: Translating Understanding into Action

Organisational and Personal Action Planning: What, When, and How?

Successfully tackling the global challenges will require everyone to take part. Clearly, each person or organisation can only influence certain areas. However, with everyone getting involved on a global scale, real and meaningful change can happen.

Driving change is not restricted to organisations, businesses or government. Every individual has a part to play in addressing the challenges that face us on Planet Earth. For example, we can each put pressure on our governments to provide leadership in key areas. When everyone starts to make even small changes they become significant. For example, let's look at how each of us could take action to deal with the Climate Change challenge. As seen, it has been estimated that if every individual in the US used air drying instead of electric clothes drying, it could reduce the US carbon footprint by over 10 per cent. Change starts with each of us.

Figure 66   We can all take part in building a sustainable world

OBJECTIVES AND STRATEGIES

No	Objectives (What am I aiming to achieve?)	Strategies (How am I going to achieve this?)	Actions (What do I have to do to achieve this?)	Outcomes (Results)	Cost ($$)	Time-frames	Responsibilities	Status
2.0	Climate Change • Limit temperature increase to 2°C • Greenhouse gas reduction of 80% by 2050 • Return to preindustrial levels long term	List strategies that you can take to address your carbon footprint. • Move to a renewable energy supplier • Reduce air travel where possible, and purchase offsets where not • Move to a more energy efficient vehicle • Use outdoor clothes drying when feasible, as opposed electricity-powered internal clothes dryers		Based on UK consumption could reduce individual carbon footprint by:				
			• Research and arrange move within the next month	~ 10%	TBD	1 month	Individual	To be commenced on..
			• Use slow travel alternatives (trains, coach) where feasible. Purchase airline offset with ticket when purchasing airline ticket	~ 20%	TBD	ongoing	Individual	To be commenced.. on
			• Replace existing vehicle with more energy efficient vehicle	~ 5%	TBD	2 years	Individual	To be commenced on..
			• Use outdoor drying whenever possible	~ 10%	TBD	ongoing	Individual	To be commenced on..

The following tables provide an opportunity to identify the priority issues for your situation – be that government, business, community, household or individual.

Each individual and each organisation can start by identifying the global challenges that they can influence. These can be prioritised in terms of both urgency and importance on a scale of one to ten. Once this is completed, simply plot the answers on the following table to establish the areas in which to focus efforts. You can then use the following table to complete a simple Action Plan – focusing on the strategies and actions that you can begin being taking: how, who and by when. A simple exercise, but the beginning of the journey in understanding how everyone can contribute to developing and implementing a Plan for the Planet on a global scale.

No	Challenge	Importance	Urgency	Total	Priority	Comments
1	Stabilising Population					
2	Climate Change					
3	Sustainable Energy					
4	Sustainable Water & Food					
5	Plant Sustainability					
6	Extreme Poverty					
7	Global Health					
8	Global Education					
9	Global Peace					
10	Financing Sustainability					

Developing Your Plan for the Planet (Government, Business or Personal)

OBJECTIVES AND STRATEGIES

No	Objectives (What are we aiming to achieve?)	Strategies (How are we going to achieve this?)	Actions (What do we have to do to achieve this?)	Outcomes (Results)	Cost ($$)	Time-frames	Responsibilities	Status
1.0	Population • A peak population of 8.0 billion people by 2050 • A total sustainable world population at an appropriate living standard through natural causes.							
2.0	Climate Change • Limit temperature increase to 2°C • Greenhouse gas reduction of 80% by 2050 • Return to preindustrial levels long term.							

No	Objectives (What are we aiming to achieve?)	Strategies (How are we going to achieve this?)	Actions (What do we have to do to achieve this?)	Outcomes (Results)	Cost ($$)	Time-frames	Responsibilities	Status
3.0	Energy Supplies • Transition to long term sustainable and non carbon based energy sources. • Maintain the 80% reduction objective in greenhouse gas emissions.							
4.0	Water and Food Supplies • Establish long term water security at national and International levels • To halve the proportion of people who suffer from hunger by 2015 • Establish long term sustainable food supplies.							
5.0	Planet Sustainability • Reduce the global deforestation rate to zero by 2020 • Reduction in number of endangered species to zero by 2020.							
6.0	Extreme Poverty • Achieve poverty reduction programs in key countries. • Fix 'leaky pipes' • Effective financial deployment.							

No	Objectives (What are we aiming to achieve?)	Strategies (How are we going to achieve this?)	Actions (What do we have to do to achieve this?)	Outcomes (Results)	Cost ($$)	Time-frames	Responsibilities	Status
7.0	Global Health • Ensure safe and reliable water supplies • Eradiate major infectious and lifestyle disease.							
8.0	Global Education • Increase global literacy and numeracy levels • Ensure 70–80 million children now enrolled in schools are able to attend.							
9.0	Global Peace • Strengthen the UN Peacekeeping capabilities. • Strive for the elimination of weapons of mass destruction • Minimise global conflicts for 2010 to 2030 transition period.							
10.0	Financing Sustainability • Establish financial stability until 2030 • Establish global funding strategy to address the Global Challenges.							

PART IV
Managing a Plan for the Planet

'So much of what we call management consists of making it difficult for people to work together.'

Peter Drucker

Best Practice Global Management

Having reviewed the ten Global Challenges and given examples of opportunities and interactivity, the next stage is to create plans at every level. However, before doing that it is essential to take a deep breath and ask why the often apparently wonderful plans created in the past have been so disappointing.

Certainly there has been no shortage of investment, aid funds and goodwill in responding to these challenges. What is so frustrating is how ineffective they have been. Only a small proportion of aid reaches the poor and needy at the front end. New factories and community facilities are sometimes built without the funds to pay the staff long term to operate them. Corruption at every level bites into scarce resources. There is often a lack of focus on the few changes which will make the most difference. Inadequate measurement of results and tests for 'value for money' are missing as is 'customer focus'. Wasteful overlaps in the work of different agencies are commonplace. Too many top-down master plans from the western world are made without listening to those in need and understanding their unique cultures and their specific requirements.

Put simply, the major cause of plans failing is usually ineffective management. It is therefore useful at this point to look at what good management is.

Management is often confused with bureaucracy, administration or simplistic command and control structures. However, effective management is the process of leadership, direction and control, which has over the last 50 years been developed on a global scale by business enterprises. It is a generic process that performs the same fundamental tasks for every type of institution regardless of country, culture, public or private sector, type of industry or service. It enables everyone in an organisation to work together with a common Vision to achieve results. This is why the application of effective management is so important when we look at implementing our Plan for the Planet.

To achieve this, it is useful to study the ten best management practices of successful business. Then each manager can review these against his or her own organisation and role, using a series of effective management 'health check' checklists which follow in this section.

Finally, the adjustments which need to made be when applying business practices to the public sector will be considered.

Let's start with a review of the ten global management best practices:

1. Effective Business Planning
2. A Customer-Driven Organisation
3. Sensitivity to the Outside Environment
4. Thinking Global: Acting Local – Global Partnerships and Teaming
5. Understanding Competition
6. The 80:20 Rule – Focus
7. Productivity and Continuous Improvement
8. Mastery of Information Technology
9. Shared Values
10. Effective Change Management.

CHAPTER 16
Ten Global Management Best Practices

1. Effective Business Planning

Few would deny that effective planning is at the heart of good management. Without a clear Vision, objectives and strategies, internal confusion is inevitable and long-term success is unlikely. This is true for any organisation, whether it is public or private, global, national or local.

The planning process itself is deceptively simple: establish a clear Vision and objectives. Analyse the key functions and activities of the organisation and test them against this Vision and objectives plus changes which are occurring in the external environment. Agree unit, functional and individual objectives with financial and measurement criteria. Information systems are developed which track performance against expectations. Most managers consider this process as straightforward common sense.

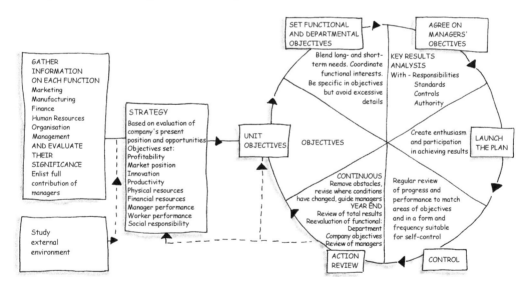

Figure 67 The Business Planning Cycle

However, successful businesses know that to do these tasks well requires focus and discipline to avoid common failures such as:

- **Objectives that are not coherent**: When Neville Isdell took over as CEO of Coca-Cola in 2004 the sales of sparkling drinks were in decline and there was a simmering war going on with independent bottlers on price increases. There were, within the company, several strategies and plans but, as Mr Isdell said, 'There was no coherence in them and there were clearly mixed messages out there.' He involved 150 of the company's top managers, identified what was fundamentally wrong and what was required. The outcome was a global business plan. Since then the company has recorded a steady international sales growth of more than 4 per cent per annum for the last 12 quarters.[1]

- **Objectives which are not ambitious and challenging**: This is particularly true of companies that have enjoyed a long period of success. The temptation to keep doing the same things but 'just a bit better' is hard to resist. This is a time for the leaders to compel a radical review, challenging literally every aspect of the business, shaking complacency and agreeing genuinely ambitious targets. Businesses, in a sense, need to constantly reinvent themselves. IBM, for example, has switched over time from being primarily a manufacturer of computers and related equipment to being one of the largest and most successful service companies, with high profits from outsourced contracts and consultancy. Effective business planners understand the changes in the environment and are able to anticipate their impact and adapt to the changes.
- **Too much detail**: Trying to plan in detail for every aspect of the business makes it difficult to focus on the relatively few objectives and strategies which have exceptional leverage for success. Also, in a fast-changing world there are unexpected crises such as the 2008/9 credit crunch and unexpected opportunities when, for example, a major competitor becomes bankrupt. Swift adjustment and change is only possible where the key success factors are simply stated and not obscured by detail.
- **Failure to understand changes in the outside environment**: The greatest challenges and greatest opportunities do not lie within the business itself but by responding to the major forces which are transforming the world in which it operates. This is particularly relevant when looking at a Plan for the Planet.
- **Reluctance to ask tough questions**: 'What should we stop doing?' is not a popular question since managers are quick to defend their territory and job security. However, every organisation inherits the unfulfilled promises of yesterday's plans: the tail end of the product line which no longer contributes; customers who seemed to have great prospects but which were never realised; an acquisition which seemed right at the time but in practice does not fit the present strategy.

Successful business plans have the courage to identify such debilitating legacies and get rid of them swiftly. They constantly ask questions such as:

- What are our core competences? How can our organisation build on these to consistently offer a unique proposition to our market which is better than our competitors?
- Are there activities which we do not do well and which should be outsourced or closed down?
- Is our record of innovation acceptable? What percentages of our products or services are new in the last three years?
- Is it timely to make a radical study of our pricing policies?
- Do we listen – really listen – to our stakeholders, investors, customers, staff and end users?

---

1 As reported in the *Financial Times*, Final encore for a man of the people, London, 9 June, 2008.

– How does our return on equity and performance compare with industry average and our fiercest competitors?

  This is not a comprehensive list but an indication of the tough questions to be answered before the plan is agreed.
- **Effective governance**: A good business plan is useless if it is not well managed and implemented across the organisation. The danger is that once developed the plan will be 'filed and forgotten'. Successful businesses don't work this way. They use the global plan as a tool to mobilise their people and their resources to achieve their objectives. This requires effective governance – overall management of the plan so that everyone understands the Vision, the objectives and strategies and their role in achieving success.

Organisations that lack good long-term planning tend to bounce from one crisis to another, often with responses that might meet a short-term need but fail to address the requirement of long-term viability. As the pressures increase, reactions become more chaotic with increasing competition for limited resources weakening the ability of the organisation to work together to address the demands being placed on it. The result is often fragmentation and failure.

These principles apply equally when looking at a Plan for the Planet. Without an overall global plan to address the challenges in a coordinated and cooperative manner it will be increasingly difficult to drive the changes and adaptations required to meet these challenges in the timeframes demanded.

As we have seen, the challenges facing us on Planet Earth are the most significant in the history of its civilisation. The human race cannot successfully move forward without a global plan – effectively managed and implemented at both global and local levels. Good management principles in this area provide a quickly replicatable approach to enable us to develop and deploy our Plan for the Planet.

2. A Customer-Driven Organisation

The most important element in developing an effective business plan is to see it through the eyes of the customer or end user. This is equally applicable, whether the plan is for business, government or a Plan for the Planet.

The real asset of a business is its ability to secure, satisfy and retain customers better than its competitors. Profits are never made within a business but only through satisfied customers. This is obviously the major difference between business and the public sector where very often the citizen has no choice but to use the services of a state monopoly. Dissatisfied business customers can vote with their feet. In the democratic public sector, citizens mobilise to replace governments that do not deliver. In the end, it is in the interests of all organisations to seek to achieve satisfied customers.

Successful businesses acknowledge that it is neither wise nor possible to serve everyone. This means a thorough and objective analysis of the business objective and therefore the market segment to be targeted. For example, is the strategy to be a mass market clothing retail business or a group of specialist boutiques? It calls for a deep understanding of what selected customers requirements are and what they perceive as value.

194 Developing a Plan for the Planet

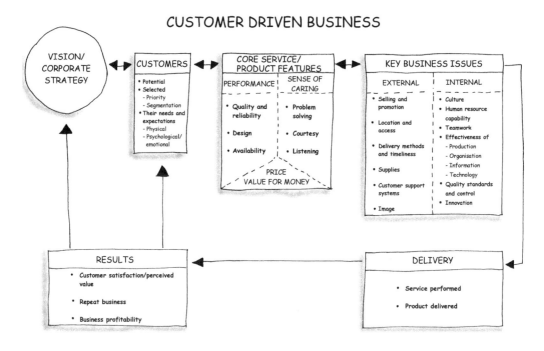

Figure 68   The Customer-Driven Business

A common strategic error is to think that price is invariably the major purchasing determinant. Factors such as brand reputation, quality, location, level of customer service and ongoing support are all important incentives to purchasing. Extraordinary guarantees that do not merely limit a customer's risk and are a commitment to uncompromising quality are essential features of a customer-driven plan. Every complaint is an opportunity to learn about quality problems. Every warranty claim is an opportunity to convert a dissatisfied customer into a loyal one. Again, these basic principles apply to all organisations – public or private.

L.L.Bean, in Freeport, Maine has a prominent guarantee in its catalogues and stores. It states simply: 'Everything we sell is backed by 100 per cent unconditional guarantee. We do not want you to have anything from L.L.Bean that is not completely satisfactory. Return anything you buy from us at any time for any reason it proves otherwise.'

That level of unconditional commitment in their business plan was made in the belief that their quality and service was so good that there would be few claims and that the guarantee would increase sales by giving customers a sense of confidence. Both these beliefs have proved to be correct over many years.

Another example of thoughtful customer service planning is a high-tech company which offers free technical advice and problem solving for any product purchased.

Thinking through every aspect of the business plan from a customer or end user point of view creates the focus required to increase efficiencies and to build a loyal customer base in business – and satisfied end users in service organisations. This applies equally when we look at the development of specific elements of our Plan for the Planet.

3. Sensitivity to the Outside Environment

From the moment a plan is approved and implemented it is subject to changes in the turbulent environment in which it operates. The dynamic changes to our understanding and impact of the challenges such as Climate Change demonstrate that this is not only applicable to business but also to any plan on sustainability. The ability to identify these changes and respond swiftly is a prerequisite for a sustainable business and planet. This is therefore clearly true of any business plan, but most importantly for a Plan for the Planet.

Relevant business examples are commonplace. The traditional Swiss watch industry had a major set back when it misjudged the importance of quartz technology. It has successfully responded by adjusting to these changes.

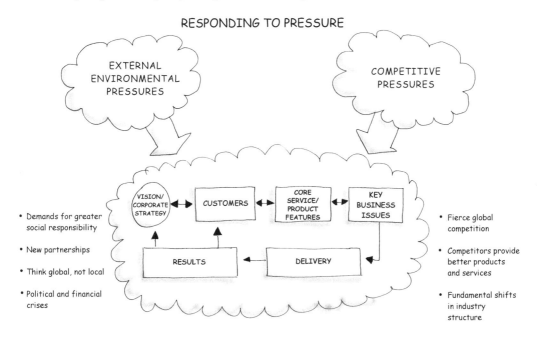

Figure 69   Responding to external pressures

European shoe businesses, which had outsourced their production to the Far East, suffered bad public relations and lower sales when NGOs revealed that there were children in the overseas plants working in unhealthy environments and being badly paid.

Some manufacturers perceived the enormous growth in the Chinese economy as a competitive threat since good quality was produced at substantially lower costs. With the same facts, insightful companies adjusted their plans and set up joint ventures in China, and due to their early entry, are now in a strong competitive position. This underlines the importance of collecting information about the outside world and using this information to develop effective plans and strategies.

Technologies are changing at an exponential pace in every industry: from valve to transistor to microprocessor in half a generation. Those slow to respond are no longer in business. Fixed-line communications systems are struggling as global markets in mobile phones have been created in a very short time span. Farsighted car manufacturers, recognising the implications of high oil prices and the imperative to have a sustainable society, are producing 'green' hybrid cars. A car manufacturer that fails to move quickly in this field will be bankrupt in a decade.

Successful businesses anticipate changes in the way business is viewed in society. They understand that profitable survival and sustainability increasingly depend on partnerships with NGOs and governments in finding answers to the Global Challenges of water and food shortages, Climate Change and poverty.

When deciding to make a new overseas investment, a thorough study of the political and social context is now as important as the traditional economic analysis of market opportunities.

A few years ago many companies treated Climate Change and its implications for energy as interesting but not vital in their planning process. Once again, a company that is not measuring its carbon footprint, conserving energy and designing products and service that fail to make a positive contribution to this and the other Global Challenges will not prosper.

Successful companies systematically collect information on external trends which are important for their business. Beyond that, senior managers can spend more time working with organisations such as the World Economic Forum or the UN Millennium Development Goals. First-hand contact with opinion formers outside business as well as business leaders from other countries and industries is a necessary time investment to understand global changes at an early stage.

The challenges present an opportunity to business to deliver not only to their immediate stakeholders – shareholders, customers and staff – but also to the broader community of stakeholders in achieving a sustainable Planet Earth.

The pioneering work of the worldwide network of research and development scientists and researchers is constantly creating new tools and solutions to respond to the Global Challenges.

4. Thinking Global/Acting Local – Global Partnerships and Teaming

The interconnectivity of the world economy is a fact of business life and has implications for medium-sized companies as well as for the large international organisations and, as we have seen, for a Plan for the Planet.

An effective business plan has to take into consideration threats such as low-cost competition from fast-developing nations such as India, China and Brazil and new opportunities such as attracting major foreign investors and increasing exports.

There is a complex balance needed to capitalise on the opportunities. A company must define the strategic areas that it must control centrally and those which can be delegated. Research and development policy needs central direction even if the laboratories and research and development centres are scattered round the world. The plan has to consider the total financial strategy of the business. Hedging against currency fluctuations and making sure a subsidiary company is not borrowing money when there are excess funds in another subsidiary are examples of this. An overall plan is essential to manage outstanding talent within the business and organise the career development and promotions of this scarce resource. This is equally relevant in a Plan for the Planet.

Think global: The general principle followed by successful businesses is to retain at the centre only those decisions, facilities and coordination that cannot be delegated without risk to the overall health of the business. This approach contrasts with the practice two decades ago where all real power was held at the centre and tight control and direction was exercised over the subsidiary and regional divisions and companies. This lends itself well to the model already being used where global objectives are developed by groups such as the UN for the Millennium Development Goals which are then implemented at a national level, adjusting for local and regional requirements. The EU has taken a similar approach in agreeing, setting and measuring greenhouse gas targets for its member countries.

However, the trend towards more decentralisation needs to be managed in response to changing global and regional requirements. Ernst and Young, one of the world's largest accountancy and service groups, has shifted from a loose federation of national entities to five regions each run by a single management team. This is the only way to provide multinational clients with the consistent level of service they need.[2] It is important in the implementation of any global plan to constantly audit the management performance of subsidiary companies and organisations.

Act local: To build strong market share in a world where expectations and national preferences are much stronger, countries must fit in with local requirements, culture and tastes. In business, a major soup manufacturer may have a global brand tomato soup but minor adjustments in flavour will be made for different markets. Successful global businesses no longer staff their overseas subsidiaries mainly with expatriate senior managers but have developed competent local managers who have language fluency and really understand the local economic and political situation. The main board members of multinational corporations have a much higher proportion of non-home country nationals than was the case in the past. At the highest level, input is called for from those who understand the local scene is part of the controlling board.

Global teaming and partnerships: The underlying values of trust and partnership that govern behaviour are easier to experience than define. However, they are playing an increasingly essential role in successful businesses and in building a sustainable planet. The importance of this global teaming is evidenced in the partnership between global scientists in developing our understanding of Climate Change and its damaging impacts.

Global policies on Climate Change can only be seriously influenced when there is collaboration between business, NGOs and government with up-to-date information from the 'front line'. Front line information on poverty calls for the needs of local organisations to be met in partnership with the parent company. Networking and partnership by all companies in an industry sector are essential and the commercial risks of more transparency are small compared with the benefits.

Another area for close cooperation is innovation where customers and suppliers pool their knowledge to solve problems common to them both. This may seem self-evident when historically there has been a tense, even antagonistic, relationship between the two. Jim Owen, Chairman and CEO of Caterpillar Inc. says of their independent dealer

---

[2] As reported in *The Sunday Times*, Running through the numbers, Andrew Davidson Interview, London, 31, February 2008.

198 Developing a Plan for the Planet

network, 'They're not our customers. They're our partners in delivering value to end-use customers of our products.'

More than half of the CEOs surveyed globally in 2008 by Pricewaterhousecoopers think that collaborative networks will be a defining organisational principle for business. The purpose of participation in such networks include: sharing best practice, creating innovation, influencing policy and reducing costs.[3]

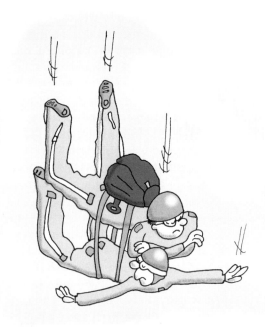

Figure 70   I think it's your turn to pull the ripcord! No, I think it's your turn!

The importance of these best practices in developing and achieving a successful plan for Planet Earth cannot be underestimated. The ability to coordinate at a global level, deliver at a local level, and to collaborate through effective teaming and partnerships will be one of the key success factors in tackling the Global Challenges.

5. Understanding Competition

Competition has intensified greatly as part of economic globalisation. Huge sectors of the world's economies have been deregulated and privatisation of State-held enterprises has accelerated. Traditional industry definitions have changed the nature of competition. For example, many food supermarket chains now derive substantial income from non-food items such as clothing and electrical goods. Capitalising on their customer loyalty and brand strength, they are moving into financial services. This raises new challenges for sectors such as clothing, electrical goods and financial institutions. Stability and market dominance are rare characteristics these days.

---

3   As reported in Compete and Collaborate, Pricewaterhousecooper 11th Annual Global Survey, 2008.

Effective organisations understand that their long-term success and sustainability depends on securing and holding customers better than the competitors and therefore a plan that is not profound in its analysis of competition will fail. Although competition is often seen as a negative influence it is the catalyst for change and improvement on which survival depends. It is a key reason why business is often more efficient than public services.

Typical questions raised are:

- What is our market and customer segment and what is its current size and growth rate?
- How cyclical and seasonal is the industry and how can we get a better balance?
- Should we seek to acquire competitors or collaborate with them? Or is it in the best interest of our investors to sell our company?
- Are there fundamental shifts in the sales, distribution and service delivery channels?
- Are we threatened by substitute products/services?
- Are there significant technological changes which will influence our business success?
- Is the bargaining power between ourselves, our customers and our suppliers changing?
- For each of our major competitors, how do we compare in terms of market share, profitability, product/service innovation/growth, commitment to risk and entrepreneurial culture?
- What do customers perceive as our distinctive competence, contribution and value for money compared with our competitors? What is our unique selling proposition?
- Are we concentrating too much on improving our competitive operational effectiveness and not paying enough attention to our fundamental strategic positioning?
- Are we competing with organisations for resources when we could be collaborating with them, for example, in the case of government departments and NGOs?
- How easy is it for new entrants to secure a place in our industry?
- Do we benchmark ourselves against best practice in our industry?

IKEA has positioned itself primarily as a supplier of stylishly designed, ready to assemble, furniture at a low cost made possible by little reliance on outside manufacturers. The ranges are simply displayed in the store and customers make a purchase and pick up their product from the warehouse. Many competing traditional furniture businesses require orders to be placed and there may be weeks before delivery. Creative strategies to effectively address competition are at the heart of good business planning.

There is increasingly an understanding of the principle of cooperation/competition – of where competition and cooperation can go hand in hand. Shared R&D in the development of solar energy and other sustainable energy sources, for instance, is a key opportunity to fast track the development of these technologies in the short timeframes required to bring it to market. Successfully achieving our Plan for the Planet will require new and innovative ways of looking at competition as well as cooperation to allow us to address the challenges in the limited timeframes available.

6. The 80:20 Rule – Focus

Focusing on the relatively few actions that determine success or failure is essential. Managers and employees are often bewildered by too many conflicting priorities. Flexibility and quick response to changes in the external environment is only possible where a small number of factors have to be adjusted. This will be a priority as we seek to achieve maximum impact on the Global Challenges in the required timeframes. Finances and resources, which are always limited, are unproductive if they are not concentrated on the important critical success issues.

This principle is called the Pareto Principle after the Italian economist, Vilfredo Pareto, who observed in 1906 that 80 per cent of property in Italy was owned by 20 per cent of the population. In business terms the concept was developed and popularised by J. M. Juran, a renowned American consultant in quality management. It is widely used in total quality control and six sigma programmes. Juran was always challenging his clients to distinguish between 'the vital few and the trivial many'.

Of course, the figures are not meant to be precise but are a useful rule of thumb to test the allocation of resources in a range of circumstances.

For example:

- Do a small proportion of customers produce most of sales revenue and/or profitability?
- Do most quality defects arise from a small number of recurring problems?
- Do job descriptions make clear the limited number of tasks which have high influence on job success or failure? What proportion of a manager and employees time is spent on these key tasks?

Distinguishing between vital and trivial is a way of life in good companies. In a sense, it is part of a wider culture which is constantly seeking simplicity. Flat, coherent and simple organisational structures which everyone can understand are essential. If there are many so-called coordinators it is usually a signal of poor structure. The ability to communicate with staff and customers in ways that avoid jargon and platitudes and are to the point is another example of disciplined simplicity. Focused information that is vividly and simply expressed in terms that facilitate decisions is important.

A useful way to describe this approach is the KISS Principle: 'Keep it Simple Stupid'. In common with the other best management practices it is common sense. Unfortunately, common sense in many businesses is a rare commodity.

A simple example is the debate about whether the developing countries or the developed countries should act first to address greenhouse emissions.

Obviously both should act to avoid long-term disaster but where is the most effective point to focus? Using the 80:20 rule this becomes obvious – the top ten CO_2 countries which are emitting 70 per cent of greenhouse gases must be a priority.

7. Productivity and Continuous Improvement

Although improving productivity and effectiveness has been a way of life in successful companies, changes in the outside environment, such as the global financial crisis, have made it an even higher priority. As we seek to develop and deploy best practices in our Plan for the Planet, continuous improvement will be as essential to ensure not only rapid, but also effective deployment.

Maintaining profitability, let alone increasing it, is difficult when raw materials costs are escalating. Competition is intense so it is hard to increase sales volume. Some costs can be passed on through increased prices but the opportunity to improve margins is limited in low-growth economies. Searching for productivity improvements and efficiency throughout the business is a significant way to remain profitable and successful.

Productivity is essentially a ratio between how well an organisation converts inputs (for example, labour and materials) into outputs (for example, goods and services).

Using the 80:20 rule, an analysis can be made of those factors which have a high influence on profits. Where material costs are a high proportion of final product cost, it is useful to measure every $ of material entering the production process to see how much is lost through waste. Where is the waste taking place? What are the fundamental causes? Another example: why does expensive machinery have a low ratio of running to down time? Are productivity and effectiveness failures due to poorly trained staff? This principle applies equally to both profit and non-profit organisations, and in the application and implementation of our Plan for the Planet.

The thought process is how can we secure more output for the same input? How can we secure the same output with less input? Can we secure much more output for slightly more input? Typical outcomes include reduced unit costs, improved quality and improved delivery logistics. In pure service businesses, the most important tool for improving productivity is often the effective use of modern IT.

However, whilst manufacturing businesses have well-established productivity plans, service sector planning tends to be less well developed. Service managers point out that compared with manufacturing, service is less predictable and controlled and the tools used in production planning are not always appropriate. The underlying anxiety is that driving productivity improvements may reduce costs but damage customer satisfaction. However, McKinsey and Co have found that 'by looking closely at sources of failure and finding ways to address them, product companies can reduce costs by 10 to 15 percent whilst maintaining customer satisfaction.'[4]

Effective companies also benchmark their key performance indicators against the best similar operation, practice and function within the business and, where possible, against the best in industry and establish 'best practices'. In some instances, cross-industry comparisons are also useful for comparison and the cross-fertilisation of best practices and innovation. Supply chain efficiencies are important for the provision of aid programmes as well as for business.

8. Mastery of Information Technology

No large company report would be complete without a glowing reference to the digital revolution and its implications for the business.

---

[4] As reported in *The McKinsey Quarterly*, Improving productivity in product services, Fagan, T., Harmon, E. and Lukes, T., No 1, pp. 30–39, 2006.

For once, this is not public relations. Revolution, defined as 'a far reaching and drastic change in ideas and methods' is at the heart of a remarkable transformation in IT. In less than a generation vast global resources of new knowledge and skills have been developed together with hands-on contribution and experience. It is hard to recall that 40 years ago the computer was seen as a tool to reduce payroll costs and the expectation was for a company to have a small number of computers on site. It took 27 years to reach one billion PCs and by 2015, with growth in fast-developing countries, the figure could reach 2 billion.[5]

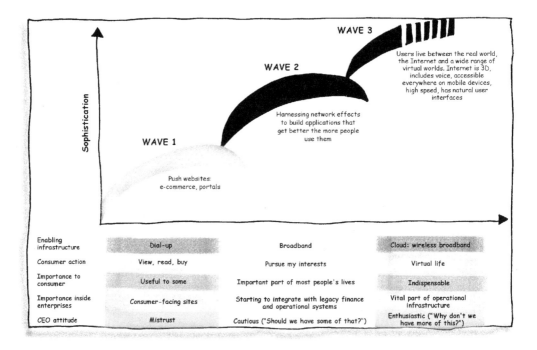

Figure 71 **The waves of Information Technology development**

The quantity of high-quality information will grow explosively as costs per transaction reduce and technological innovation extends the range of applications. Quite simply, this is the beginning of the revolution and any company which does not understand that, will fail.

Take some examples of how effective companies use IT:

- **Supply chain efficiencies**c Just In Time deliveries from supplier to customer are totally dependent on common databases and systems. Automated supply chains bring improved service delivery and lower costs. A FedEx or UPS customer can use the Internet to track the precise location of a package.
- **Partnership**c Proctor and Gamble (P&G) have for many years had a cooperative relationship with Wal-Mart who account for some 15 per cent of P&G sales.

---

5   As reported in 'One billion PCs in use by the end of 2008', Forrester Research Inc., 400, Technology Square, Cambridge, MA 02139, USA.

Originally that meant Wal-Mart giving P&G access to its customer database. The latest development is to schedule direct delivery to specific stores, cutting the progress of a product from factory to store by about ten days.[6]

- **Investment**: Capital investment decisions have so many variables it has been difficult to manipulate the range of options on which to make a wise decision. Modern IT applications and models can quickly take into consideration such factors as the expected payout over the life time of the investment and the opportunity costs versus alternative investments.
- **24/7**: An individual manager now has 24/7 access to up-to-date information through a PC, laptop, mobile phone and other innovative and emerging communications technologies.

These are indicative examples: each organisation has its special needs and solutions.

Beyond the contribution of IT to the existing operations, new entrepreneurial businesses are being created, substantially based on information. Amazon's business model uses online ordering based on up-to-date product descriptions. It can make personal recommendations by analysing every customer's past buying patterns. The warehouses are automated and, due to this high-tech infrastucture, have also been able to reduce costs and therefore offer very competitive prices. The result is that Amazon has been able to capture a major share of selected markets. Recently they have been able to extend this model by selling other retailers the right to use Amazon logistics and delivery systems.

No modern architectural practice could survive without highly developed IT systems for design and calculations. Real estate agencies traditionally obtained their business through mailings, advertisements and shop displays. They now use websites which enable a prospective customer to profile their exact needs – type of property, location, price and so on. It is possible to have a video 'walk round the property'.

In summary, successful businesses know that IT is a unique, powerful, strategic, competitive resource which can improve the productivity and profitability of all aspects of the business. Moreover, it creates opportunities for new ventures founded essentially on information.

To realise these benefits calls for highly effective IT systems and governance. There is a residual legacy where operational managers think that IT managers are out of touch with commercial realities. In turn, IT specialists are frustrated by line managers' IT illiteracy.

In some companies it is the principal driving force of the business. In others IT remains a support function but since its contribution embraces not only all internal aspects of the business but also links to customers and suppliers, it must be fully integrated. Failure to align it with the business strategy and processes can be very costly.

The road to trust between Chief Information Officer and a business unit is illustrated in the following diagram reproduced with permission from *The McKinsey Quarterly* by McKinsey and Company. The article is 'What IT Leaders Do' and the authors are Eric Monnoyer and Paul Wilmott (2005).

---

[6] As reported in the *Financial Times*, How P&G's logistics revolution supplied a new demand, London, 22 August 2008,

The road to trust

	Communication - from CIO to business unit	Understanding - between CIO & business unit	Trust - between CIO & business unit
	Overview	**Equals**	**Partnership**
CIO's focus	• Ensure transparency through extensive reporting • Seek to determine mode of IT governance	• Explain information systems by speaking language of the business unit (finance, operations) • Clarify/share problems linked to IT management • Establish personal relationships with business unit leaders	• Incorporate IT into governance of business unit • Add value in decision making across the board • Ensure that impact of IT can be measured in terms of business impact
Reaction of business units	• Do not understand content, form of CIO's messages • Perceive CIO as a superior and behave accordingly • Do not anticipate difficulties that arise, blame CIO	• Increasingly aware of problems in IT management • Comfortable making IT-related decisions • Increasingly willing to address IT-management issues, overcome misunderstanding	• Take ownership of IT decision-making process • Experience heightened sense of effectiveness, personal development

Figure 72   Trust between Chief Information Officer and Business Units

At the same time it is important to highlight that effective IT is about good information management. One of the downsides of the information revolution is data overload. It is important that the information produced is appropriate. Multiple useless reports and data can confuse the issue rather than clarify it. It's hard to see the wood for the trees. For this reason, it is important to always refer back to the overall plan when developing information technologies to ensure that the information and reports produced provide effective feedback for customers, employees and management teams and are not a distraction.

Figure 73   All I asked was did the customer like the product!

The application of this to global planning is clear. Well-placed, timely and concise information delivered through effective IT systems are essential for good management, whether an organisation is private or public.

9. Shared Values

Some CEOs identify company values as public relations or soft, motherhood statements that are not really a key for success. However, the best companies have a radically different perception and understand that strategy and values complement one another.

The behaviour of each employee is driven by a few core beliefs and convictions. What is believed – really believed, not lip service – governs behaviour. Indeed an organisation constantly under change needs a set of genuinely shared values as the glue which maintains common purpose.

Trust is essential. The general manager who speaks about the business commitment to reducing poverty in a poor country whilst being willing to buy from suppliers with questionable working condition creates cynicism. Vision, values and strategies must fit together.

The radical transformation of business through the information revolution and the increased understanding of corporate social responsibility have further driven the need for clear and values. All organisations, whether public or private, are under increasing scrutiny. An organisation's values are becoming increasingly transparent to both customers and the community.

Table 10   Typical relationships between values, vision and strategy

Values	Fit with Vision and strategy	Capacity and will to adapt	Probable outcome
Strong Shared Values	Close Fit Clarity and commitment	High	• Positive contribution to economic success • High morale and 'ownership' • Quick response to crisis
Weak Confused Values	Poor Fit Usually confused Vision also	Neutral	• Organisation drifts • Poor motivation • Uncertainty and ambiguity
Strong Shared Values	Very Poor Fit Substantial gap between new Vision and strategies and old values	Poor	• Potential economic disaster • Bitter internal conflict between existing value holders and those pressing for changes • Anxiety

An Internet blog posted by dissatisfied customers can be accessed immediately by millions of consumers. Lack of openness and honesty is quickly revealed. Therefore, clear values and effective implementation are not only important in getting everybody in an organisation delivering on organisation objectives but are also good for customer relations and profitable business. Fully committed, shared values that match strategy give competitive advantage.

Application of this principle is an integral element in the success of many companies. The Body Shop, Marks & Spencer and Orange have all built highly-successful business operations on a set of clear, well-articulated and well-executed values: values that are important for their customers and therefore not only create business opportunities, but can also build a greater sense of partnership between businesses and their customers. Take the example of The Body Shop:

> **THE BODY SHOP**
>
> The Body Shop International plc is a global manufacturer and retailer of naturally inspired, ethically produced beauty and cosmetics products. Founded in the UK in 1976 by Dame Anita Roddick, they now have over 2,500 stores in more than 60 countries, with a range of over 1,200 products, all animal cruelty free, and many with fairly traded natural ingredients.
>
> They were the first international cosmetics brand to be awarded the Humane Cosmetics Standard for their Against Animal Testing policy. And they have their own Fair Trade programme called Community Trade, making them the only cosmetics company with such an extensive commitment to trading fairly. Community Trade now works with 310 suppliers in 24 countries, providing over 25,000 people across the globe with essential income to build their futures.
>
> The Body Shop is a leader in the trend towards greater corporate transparency, and they have been a force for positive social and environmental change through lobbying and campaigning programmes around their five core values: Support Community Trade, Defend Human Rights, Against Animal Testing, Activate Self-Esteem, and Protect Our Planet.
>
> They also have their own charity, The Body Shop Foundation. Launched in 1990, they give financial support to pioneering, frontline organisations that otherwise have little hope of conventional funding. The Foundation's focus is to assist those working to achieve progress in the areas of human and civil rights, environmental and animal protection.
>
> They are part of the L'Oreal family.
>
> Source: Reproduced with the kind permission of the Body Shop International plc.

As we have seen, this is a critical part of establishing our Plan for the Planet. Success will be achieved by understanding that the things that bind us together are much more important than the things that pull us apart.

10. Effective Change Management

Adapting to change is a way of life in business. What is strategically different today is the extraordinary increase in the pace and scale of change. New global markets and new forms of competition; new materials; dramatic shifts in information resources; rising social expectations placed on business; currency and financial fluctuations; political uncertainties; and responding to the needs of a professional knowledge-based workforce. The list will differ from company to company but the challenge remains the same.

By definition a company's Vision, long-term objectives, strategies and values are not easily changed. It is their enduring homogeneous qualities which provide a secure framework in which people can confidently experiment and innovate. The temptation is to have a policy of steady incremental improvement at an easily managed pace of change.

When turbulent demands are placed on a company, the response to which may well be survival or non-survival, the only option is transformational change. The urgent need is to adapt to change and for CEOs to see cultural and operational agility as a survival skill. The Pricewaterhousecoopers 2008[7] survey of 1,100 global CEOs showed that 76 per cent of them perceived the ability to adapt to change as one of the most important sources of competitive advantage. A useful approach to effective change management is:

- **Clarify why major change is crucial**: Serious decline in profitability? A new opportunity which calls for a different set of skills and mind set? HP, for example, responded to IBM's growth in outsourced and consulting business by acquisition and changing their internal organisation. A telephone business with an expensive fixed-line legacy has to change its business model quickly or it will become bankrupt. It is important to define precisely the output required from the proposed radical change. This is particularly relevant to any global plan, whether that is a business plan or a Plan for the Planet. People, as customers, employees, investors or other stakeholders need to understand why change is required and that they can make a difference.
- **Executive endorsement**: Nothing happens without the total commitment of the top management team who must articulate the need and plans, be role models of leadership and enthusiastic supporters of the changes. Often this calls for a CEO with a maverick reputation, such as Jack Welch's time at the General Electric Company, and sometimes it is necessary to bring in an outside person. With a Plan for the Planet this will require the alignment of leaders of business, government and community to support the key initiatives required to achieve long-term sustainability. Too often, top managers want to change others but not themselves.
- **Effective programme management**: A rigorous change plan with a Vision, objectives, strategies, objectives resources and timescales must concentrate on the key changes which will give high leverage for success – with focus on the delivery of the end benefits for key stakeholders and end users. Change often demands new skills and capabilities so that learning programmes will be required. Programme management provides an industry best practice approach to coordinating the key resources to achieve these objectives. Again, these principles can and should be applied in both private and public organisations.
- **Involvement by everyone**: If the message is clear and convincing then most people will wish to make a contribution. Top-down plans alone do not succeed. People have to feel ownership of their own changes. Marks & Spencer focused on employee education and involvement when rolling out their corporate responsibility Plan A. Success at a customer level was only going to be achieved if everyone in the organisation understood the Plan A approach and was committed to its success.

---

7 Available at: pwc.com/ceosurvey.

It is also important to assign outstanding mangers to major 'change agent' roles. The potential sources of resistance to change should be identified and plans developed to respond to the resistance. The use of global task forces and work teams to address complex issues is an important best practice which brings together groups of people chosen for their knowledge and experience, not their organisational position.

Where possible, identification of and early focus on 'easy wins' gives encouragement and demonstrates quick progress. Certainly success of all kinds should be celebrated. Major changes invariably require new knowledge and skills and that calls for significant investment in education and training.

Table 11   Characteristics of static vs future-oriented organisations[a]

Dimensions	Characteristics	
	Static Organisations	Future-Oriented Organisations
Structure	Rigid: permanent committees, reverence for constitution and bylaws, tradition	Flexible: temporary task force, readiness to change constitution and bylaws, depart from tradition
	Hierarchical: chain of command	Linking: function collaboration
	Role definitions: narrow	Role definitions: broad
	Property: bound and restricted	Property: mobile and regional
Atmosphere	Internally competitive	Goal-oriented
	Task-centred: reserved	People-centred: caring
	Cold, formal: aloof	Warm, informal: intimate
Management and Philosophy	Controlling: coercive power	Releasing: supportive power
	Cautious: low risk	Experimental: high risk
	Errors to be prevented	Errors to be learned from
	Emphasis on personnel selection	Emphasis on personal development
	Self-sufficient: closed system resources	Interdependent: open system resources
	Emphasis on conserving resources	Emphasis on developing and using resources
	Low tolerance for ambiguity	High tolerance for ambiguity
Decision Making and Policy Making	High participation at top, low at bottom	Relevant participation by all those affected
	Clear distinction between policy making and execution	Collaborative policy making and execution
	Decision making by legal mechanism	Decision making by problem solving
	Decisions treated as final	Decisions treated as hypotheses to be tested
Communication	Restricted flow: constipated	Open flow: easy access
	One-way: downward	Two-way: upward and downward
	Feelings: repressed or hidden	Feeling: expressed

a   As reported in OPTIMUM, Characteristics of static versus future oriented organisation (Organisations of the future), Lippitt, G., Vol, 5, No 1, 1974.

- **Performance tracking**: As with any good plan, rigorous tracking of progress, measuring results against objectives, is necessary. This information must be communicated to

everyone as the change process takes place. Celebrating milestone success is also important in building motivation and commitment to the change and the change process.
- **Think long term**: To expect a major transformational change plan to be successful in a short time is unrealistic, like expecting an oil tanker to change direction quickly. However, there are usually significant gains to be made step by step throughout the implementation of the plan and securing quick wins.

Failure to respond to challenges and opportunities which can only be met by transformational changes in the organisation threatens the survival of the business. The task of the CEO and top management team is to integrate the head with its logic and analysis – and the heart –with its emotion and excitement.

These change management principles apply particularly to our Plan for the Planet. Establishing for everyone a clear understanding of the issues and their interconnectivity, as well as the Vision, objectives and strategies to address them – provides the framework for global involvement and action.

CHAPTER 17
Global Management Best Practices: Applications to Managing a Plan for the Planet

Application of Global Management Best Practices to Public Sector Management

The business contribution in addressing Global Challenges is important. Equally important is the contribution made by the public sector, where a high proportion of a nation's gross national product is spent. This sector has to perform well to deliver the essential needs of its citizens in healthcare, defence, education and the like, even where there is some partnership with business. In addition, government sets the framework of law and regulation which governs the way business itself can operate.

One reason why national governments in the developed nations don't give priority to the wider global threats is because they are preoccupied with their own country problems. This is where global teaming between countries can be highly productive, by developing and transferring skills and best practice from one country to another.

Internal problems include demographic structural change which will swell the ranks of the elderly to about 20 per cent of the population in the US and 30 per cent in Europe and Japan in the next two decades. The burden of pensions and healthcare will be formidable. Raising taxes on a reduced working population is politically risky. Cutting service levels by cost reduction only reinforces the increasing disillusionment of citizens with the political leadership.

Improving productivity is essential for grappling with these problems even though measurement may be less quantifiable. Since productivity levels are often lower than the equivalent in the private sector, the application of best practice management principles can be of great assistance to the public sector. For example, the UK Office for National Statistics states that the average public sector working output in 2007 was 3.2 per cent lower than in 1998. Over the same period, market sector productivity rose by 22.8 per cent.[1]

However, the temptation to directly apply business management models must be tempered with the constraints of public service requirements. Obviously there are important key elements in common such as defining the purpose of an organisation, ensuring key stakeholder requirements are met, setting objectives and developing strategies and resource allocation. However, there are also crucial differences:

---

1   As reported in *The Sunday Times*, London, 14 June, 2009.

- **Customer focus**: Citizens and end users are not necessarily the same as business customers. A government is obliged to give universal service and rights. Government cannot ignore a costly minority. Citizens often have no alternative to state provided services. However, the emphasis on customer and end user focus is becoming increasingly recognised as a key to success in government organisations.
- **Political timeframes**: Many of the strategic decisions such as nuclear versus coal-fired power plants need a long time horizon which does not fit comfortably with short time political cycles.

    There are abrupt and costly changes made when new governments change priorities. Decisions are at times made on grounds of political expediency rather than the objective, most productive grounds. Management is sometimes top-down, bureaucratic and resistant to change.
- **Financial management**: Businesses are paid for satisfying their customers. Public sector institutions typically get paid out of a budget allocation that often has historically had unclear links with end user satisfaction. Daily experience demonstrates that putting the customer (the end user) first is not always evident in the public sector, however, increasing emphasis is being placed in this area.
- **Objective setting**: Measurements are difficult and arbitrary targets, at times driven by political agendas, can distort actual operational effectiveness. Increasingly however, customer-driven target setting is being introduced with significant results.
- **Managing competition**: Competition between institutions and government departments for budget allocation often damages effectiveness. This is at times seen with NGOs, where cooperation could be beneficial. Even in the developed world we see this in health systems. For example, a frail elderly person can continue to live at home with three half-day support visits a week. Without this, the person goes into hospital or residential home at many times the cost. Social services, protecting their budget, will sometimes be willing to see the person go into hospital, which is the responsibility of the health system. Here competition between departments can be detrimental to overall effectiveness.
- **Responsibility**: Accountability for failure is often low. The politicians who authorised a major project which is unsuccessful and those who managed it inadequately usually survive. In business it is much more likely they would not.
- **Continuous improvement**: Above all – and it is inevitable – business management is driven by the threat and discipline of the market: failure to secure and hold customers means going out of business. The public sector has different pressures, often political, which have not always fostered continuous improvement – however, as these effective business management principles become more firmly established, they are being increasingly deployed in the public sector with successful results.

Achieving Best Practice

As highlighted, over the last 20 years, public sector leaders are increasingly aware of the need for increasing productivity and there are pockets of excellence and encouraging initiatives and best practice.

In the UK, for example, public servants have been responding to the efficiency drive designed by Sir Peter Gershon. The original target was 2.5 per cent savings per annum in the three years to 2008. This has been raised to 3 per cent for the next three years. This is in addition to labour force reductions.

Savings claimed in the first three years amounted to £23 billion. Achievements include a shared personnel system for all three armed services and procurement improvements in many departments. Most departments now have a qualified finance director when in 2004 most did not. Public/private initiatives bring in business experience.

These are just examples but they highlight that there is an important opportunity for public sector institutions to dramatically improve their performance management.

So, how can these effective management principles be used for the implementation of a Plan for the Planet in the public sector? In essence, three things are necessary:

- **Total political commitment**: When shared by all political parties this can transform the management performance of the public sector and contribute to the Plan for the Planet objectives. The approach used in the UK in the Second World War of a war cabinet, devoid of political party self-interest and focused on achieving victory, is a useful best practice model. The global coordination to address the 2008/9 global financial crisis provides a more recent useful example of greater global cooperation and coordination.
- **Customer focus**: A change of focus to a customer (end user)-driven obsession for every strategy, investment and process. Understanding who the 'customers' are and driving to meet their requirements can simplify decision making and priority setting as well as increasing political effectiveness and 'breaking down the silos'.
- **Continuous improvement**: Sustained, ambitious improvements in productivity in every department, every process, combined with better measurement tools and reward/punishment systems to insist on accountability.

Application of the Best Practice Global Management to International Organisations

Successfully addressing the Global Challenges requires a united global effort which integrates, and where necessary, transcends the individual self-interest of nations and states. Failure in one area will mean failure for all. Successfully addressing the challenges requires a truly global teaming effort, translated into local action. Cooperation between government, business and people at a local, national and international level is essential.

In management terms this is a challenge in itself as demonstrated by the mixed success in tackling the Global Challenges in the past. However, whatever legitimate criticism may be leveled at organisations such as the UN, World Bank, IMF and WHO, they provide an important focal point for leading, energising and coordinating the response to the Global Challenges.

Representing over 220 nation states, the UN is the only global organisation which can claim to speak for the world's common interest. It is sometimes bureaucratic; its worthy separate institutions do not always integrate and coordinate their efforts; national agendas sometimes weaken it and regional political lobbying and power plays sometimes impede the best decisions. Promotion is too often on the basis that it is 'a country's turn' rather than selecting the best qualified candidate.

The challenge is therefore to leverage the capabilities of global organizations, such as the UN, by focusing on improving management performance and enhancing the support of nation states and appropriate funding. Those who believe that this is an unrealisable dream must ask the question: What is the alternative global institution? We do not have another 50 years to further develop the UN's international governance approaches. The challenges are pressing now.

These are not academic concerns. We need to constantly remind ourselves of the harsh reality that the whole of the human race on Planet Earth is threatened now by unsustainable population growth, Climate Change, energy shortfalls, food and water crisis, and global diseases. Well-coordinated and cooperative global plans are essential for long-term sustainability.

When faced with life and death battles for survival it is necessary to put aside differences and work at all levels – international, national and local – with an urgent common purpose.

The successful implementation of plans requires competent management at every level in public and private organisations. In the next section, a Health Check is provided to allow management effectiveness to be assessed and improved in every type of organisation.

CHAPTER 18
Global Management Best Practices: A Health Check

The successful implementation of global plans to meet the ten Global Challenges requires every organisation, every manager and every individual to seek improved management effectiveness.

The following Best Practice Global Management Health Checks are tools for this purpose.

- A health check list to see how the ten Best Practice Global Management principles are being applied in an organisation.
- A series of Best Practice Global Management health checklists for each individual manager to complete.
- The check lists can also be printed off the Plan for the Planet website: www.PlanforthePlanet.co.uk.
- To use these, adjust the language and details to suit your organisation. For example, 'customer' would be appropriate for a business manager; 'citizen or end user' may be correct for a public service manager; 'HIV affected person' could be appropriate for an NGO in a developing nation working in the health sector.
- It is then important for management teams to discuss the results of the questionnaires, list improvements including benefits and who would be responsible for implementation, define the benefits and convert the results into action plans.

Application of the Ten Best Practice Global Management Principles

The following checklist provides the opportunity for each manager and individual in each organisation – be that business, government or NGO – to complete the following analysis and produce a summary of improvements and recommended action plans based on the 10 global management best practices.

	Management Principle	Present Position in my organisation (Strengths, Weaknesses, Opportunities, Threats)	Proposed Improvements
1	Effective Business Planning		
2	Customer-Driven Organisation		
3	Sensitivity to the Outside Environment		
4	Think Global: Act Local: Global Partnerships & Teaming		
5	Understanding Competition		
6	80:20 Rule		
7	Productivity & Continuous Improvement/Best Practice		
8	Mastery of Information Technology		
9	Shared Values		
10	Effective Change Management		

Building a Sustainable World – An Effective Manager's Health Check

Every manager and individual in every organisation – business, government, community group or NGO – will need to perform to the highest level if the Global Challenges are to be met. To achieve this, it is important to consider the following key success factors for management effectiveness:

- A clear organisation structure and roles.
- Effective dynamics which drive the structure.
- Effective information – its quality and relevance.
- Managing change.
- Managing innovation.
- Managing motivation.

The following detailed checklists are included to assist managers to review their current position in each of these areas, and develop action plans to enhance efficiency and effectiveness.

A. Organisation Structure and Roles

		Strongly Agree			Strongly Disagree		Don't know	SUGGESTIONS FOR IMPROVEMENTS... BE SPECIFIC
		1	2	3	4	5		
1	There is a clear statement of corporate or organisational Vision & objectives for the current year.							
2	The organisational structure is designed to put the customer first at all times.							
3	There is a clear statement of objectives for the department/function in which I work.							
4	The organisation structure is reviewed regularly to ensure that it facilitates the achievement of these objectives.							
5	There are written, clearly expressed policies on major issues. (A policy can be defined as a standing company or departmental answer, or guideline to questions which are frequently asked.)							
6	The fundamental division of work (e.g. into main functions such as development/ sales/production and operations/marketing/ human resources or by geographical division) works well. There is effective communication and coordination between these groups.							
7	The structure emphasises those areas which are central for your organisation's success.							
8	The degree of delegation of responsibility and authority is effective.							

		Strongly Agree			Strongly Disagree		Don't know	SUGGESTIONS FOR IMPROVEMENTS... BE SPECIFIC
		1	2	3	4	5		
9	The structure would work better with fewer levels, i.e. a flatter structure.							
10	The structure would be more effective if some of the large units were divided up into several small units.							
11	Closely related activities or functions are sensibly assigned to the same organisational unit.							
12	The span of control (i.e. the number of people reporting directly to one manager) is effective.							
13	Each individual manager and employee is clear about the results expected of him/her and has related standards of performance.							
14	Each manager and employee is clear about the authority he/she may exercise.							
15	The work of different departments and functions is coordinated and effective.							
16	The organisation and management structure as a whole is communicated to and understood by most managers and employees.							

The best feature of your formal organisational structure is:

The greatest weakness of your formal organisational structure is:

Additional comments:

B. Organisation Dynamics

		Strongly Agree			Strongly Disagree		Don't know	SUGGESTIONS FOR IMPROVEMENTS... BE SPECIFIC
		1	2	3	4	5		
1	The key value in our organisation is to satisfy customers/end users.							
2	I have opportunities to participate in shaping and deciding objectives at: – corporate level – departmental or functional level – my own job level.							
3	People in our organisation are committed to their objectives.							
4	We are usually consulted before major organisational changes which affect us.							
5	The system is reasonably flexible – we are not a 'work to the letter of the law' organisation.							
6	The company climate is generally friendly and supportive rather than threatening.							
7	Conflict between organisation units rarely exists.							
8	Conflict is brought into the open and used constructively to clarify issues and develop solutions.							
9	I can speak my mind openly to my management team.							
10	The best-qualified people generally get promotion.							
11	Contribution and knowledge have more influence than formal organisational status.							

		Strongly Agree			Strongly Disagree		Don't know	SUGGESTIONS FOR IMPROVEMENTS... BE SPECIFIC
		1	2	3	4	5		
12	There is a high degree of personal trust and respect among people.							
13	The organisation provides adequate means (facilities, materials, machines, finance, IT support etc.) to enable me to do a good job.							
14	Boundaries between jobs and departments are not rigidly observed; there is a spirit of helping one another as occasion demands and working as one team.							
15	The informal organisation facilitates the working of the formal structure.							
16	I feel a sense of freedom in how I achieve my job objectives.							
17	The management team is not remote: they are aware of the company's needs at grass roots level and practical level.							
18	We are too complacent – we need more commitment to secure better performance.							
19	The organisation structure helps us to make quick and good decisions.							

The best feature of our organisation dynamics is:

The greatest weakness of our organisation dynamics is:

Additional comments:

C. Management Information

		Strongly Agree			Strongly Disagree		Don't know	SUGGESTIONS FOR IMPROVEMENTS... BE SPECIFIC
		1	2	3	4	5		
1	Consider each of your personal key tasks and each of the objectives of the organisation unit for which you are responsible: I have information which is:							
	• reliable and trustworthy							
	• comprehensible in content and presentation							
	• timely (not too frequent and not too late)							
	• appropriate in detail.							
2	I have good regular information on customer satisfaction.							
3	I have good regular information on our competitors.							
4	I have good, regular information on trends and developments in the external environment that affect my organisation. (e.g. technological, political, industry environmental, legal changes).							
5	Our organisation is not over preoccupied with short-term, non productive reports.							
6	We are ready to use qualitative measures and do not try to force all information into figures.							
7	There is little scope to make our reporting systems more cost effective through editing, elimination, simplification etc.							

		Strongly Agree			Strongly Disagree		Don't know	SUGGESTIONS FOR IMPROVEMENTS... BE SPECIFIC
		1	2	3	4	5		
8	There is an open, constructive relationship between the main suppliers of information and me.							
9	Our computer-based information systems and IT technology are reliable, technically up to date and compatible with one another.							
10	We make good use of the 'management by exception' principle in our reports.							
11	We have easy access to information we require but do not receive regularly.							
12	Reports etc. are designed to stimulate action and facilitate decisions.							

D. Managing Change

Consider the last major change for which you were responsible:

1	**OBJECTIVE** A precise definition of the desired results of the change.	
2	**DEGREE OF CHOICE** What degree of choice existed? e.g. legal requirements timing with other projects	
3	**FACTORS FAVOURABLE TO THE CHANGE** Technical Administrative Human Other	
4	**FACTORS UNFAVOURABLE TO THE CHANGE** Technical Administrative Human Other	

5   THOSE AFFECTED BY THE CHANGE

PERSON/GROUP	EXPECTATIONS	FEARS & ANXIETIES	LIKELY RESPONSE	ACTION TO SECURE SUPPORT AND OVERCOME ANXIETY

6   **REVIEW OF EXPERIENCE**
When the change was complete, did you make a critical review in order to learn lessons for the future? What problems came up which were not anticipated? How could they have been better handled? What went particularly well which could be used in future changes?

7   What lessons have you learned from change management in this organisation or other organisations that you or your organisation can apply to improve future change management effectiveness?

E. Managing Innovation

	INNOVATION	YES	NO	SPECIFICALLY, WHAT WILL YOU IMPROVE IN THE NEXT 3 MONTHS?
1	Do you listen to your customers, particularly those who are innovative leaders in their industry? (i.e. many major breakthroughs in products and services come from listening to customers). What do they perceive as value? What are their future plans and needs? Do you make time to go and meet and listen to your customers? Do many of your staff to visit customers?			
2	Do you set up joint projects with customers and suppliers to find innovative ways to work more productively together?			
3	Do you study your competitors? If they are taking your potential customers there must be something you can learn from them. Ask customers who left you why this happened. Go and buy from a competitor. Was it a good experience? Was the product of equal or better quality to your own? Did you test it? Did you take it apart to see how it really works?			
4	Do you set a hard, quantifiable target for innovation? e.g. % of revenue from new products/services per annum.			
5	Do you encourage – really encourage – your staff to be innovative? 'Champion' those who bring forward ideas? Get rid of the bureaucracy that strangles innovation? Are you willing to defend your staff from 'experts'… They are not always right.			
6	Do you encourage experiments, the development of best practices and pilot schemes – and provide time and budget (officially or unofficially) to finance them on a small scale? 'Perfection is the enemy of progress.' 'Too much analysis equals paralysis.' 'Big things come from small beginnings.'			
7	Do you use multidisciplinary teams and task forces or work streams? Mixed professional experience? Mixed levels of people? Limited time? Specific problems? e.g. how can we reduce development time of this service by 20%?			
8	Do you use competition? Set two people or two work streams or task forces the same challenge? Drop in on their meetings? Facilitate progress with materials and knowledge?			

	INNOVATION	**YES**	**NO**	**SPECIFICALLY, WHAT WILL YOU IMPROVE IN THE NEXT 3 MONTHS?**
9	Do you learn from failures? If failure means humiliation and/or reduced career/earnings prospects, innovation will soon dry up. Promote innovative people, praise efforts. Persevere – few tough problems are solved at first effort. It is an iterative process.			
10	Do you set an innovative example yourself? • What have *you* innovated in the last 6 months? • Do you praise/recognise experiments, improvement? • Are you *obviously* receptive to new ideas and impatient to try them?			

F. Managing Motivation: Leadership Responsibility and Effective Teaming

		YES	NO	ACTION NOTES
1	Do you make time to 'walk about' and talk/listen/observe what is going on inside your organisation?			
2	Do you celebrate and enjoy the successes of your team, giving them full and *public* credit for their achievements?			
3	Is failure by a team member a stimulus for you to spend time coaching and counseling?			
4	Are you *sure* that members of your team are clear about the results they have to achieve and are committed to these objectives?			
5	Do you check that when you give team members responsibility for a task you also give them the necessary authority?			
6	Do you sit down at least once a quarter and review progress with your team members, review and agree objectives, listen to any problems and ideas for improvement?			
7	Are you satisfied that the training arrangements are up to date and truly effective to meet rapidly changing times?			
8	Do you spend time in a *systematic* way listening to the people in your and other teams?			
9	Are your people fully informed about the department and what is going on? (Check this by asking a random sample to describe your major product/service; to say in a sentence the top priority of the department at present; to describe in a few words why the customer buys from us.)			
10	Are career paths clearly defined and well understood?			
11	Do the payment systems (basic salary levels, differentials/bonuses etc.) encourage high motivation?			
12	Do you give time and energy priority to the choices of people for promotion? (Every promotion is a signal of the real values you stand for.)			
13	Do you rank the capacity to excite and motivate others as a very important attribute in promotion choices?			

		YES	NO	ACTION NOTES
14	Do you actively encourage work streams, task forces and mixed functional/departmental teams?			
15	Do you actively encourage experiments and pilot schemes that challenge the status quo?			
16	Do you give your people a sense of 'ownership' over their areas of responsibility – space to make their own choices?			
17	Do you have a clear and written statement on your 'people policy'?			
18	Is it easy for a team member of yours with a grievance to express this to a manager other than you?			
19	Are you sure there are no ways in which you show disregard or disrespect for your people? (Not willfully but, e.g. rules and regulations which treat all employees equally.)			

PART V
Delivering a Plan for the Planet

CHAPTER 19
Leveraging the Triangle of Change

'... humanity is facing a series of grave challenges – including climate change, loss of biological diversity, threats to water sources – that go well beyond the partisan interests of individual states. Addressing these challenges will call on all the institutional ingenuity that society can muster and will require harnessing these institutions to the broader task that these challenges represent.'

Mark Halle

'We must all hang together or most assuredly, we will hang separately.'

Benjamin Franklin

Responding effectively to the global challenges will require the involvement of everybody: government – international, national and local; business – ranging from Multinational Corporations (MNCs) to small entrepreneurial organizations; and people – as individuals, members of families, churches and local communities.

Each, of course, must have its plan but these plans have to be integrated with a wider global vision and mission. The interplay and influence between the three agents of change is illustrated in the Triangle of Change reviewed in Chapter 14.

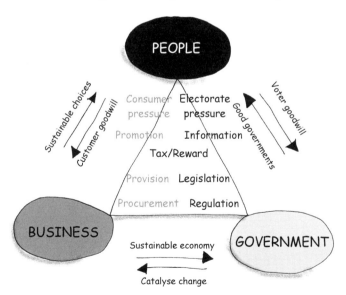

Figure 74   The key participants in the Triangle of Change
Source: Sustainable Development Commission, UK.

Leadership is essential if this process is to be successful. Leadership, however, is not confined to a few senior people but is the responsibility of each one of us. We have to commit ourselves to success with total passion; work in partnership and teams; listen to and learn from one another; constantly seek new innovative solutions; make sure good ideas are converted into practical results; monitor progress. Moreover, every leader has the responsibility to look beyond the short-term needs and objectives and see fundamental shifts in society which will shape the future.

We also need to understand how each of these agents of change is currently contributing and how their performance can be improved.

CHAPTER 20
Leveraging International Organisations and Government

1. **The United Nations**: The UN was founded in 1945 as a successor to the League of Nations. There are 192 member states. Its aims are to facilitate cooperation in international law, international security, economic development, social progress and human rights. There are over 30 important UN agencies such as WHO and United Nations Children's Fund (UNICEF).
2. **The International Monetary Fund (IMF)**: The IMF and the World Bank were established at the Bretton Woods conference in 1944 and were designed to achieve different but complementary ends. The IMF was designed to defend the international monetary system, which in effect meant helping governments to overcome balance of payments problems. Typically it works with a troubled country to agree a Letter of Intent setting out the key features of a recovery plan. It also makes long-term loans for strategic economic developments such as infrastructure – new roads and dams.
3. **The World Bank**: The World Bank is like a cooperative in which 185 member countries are shareholders. It provides financial and technical assistance to developing nations, seeking to reduce poverty through projects in education, health, infrastructure, communications and government reform. It is made up of two unique development institutions – The International Bank for Reconstruction and Development (IBRD) and the International Development Association (IDA). It also has three affiliates: International Finance Corporation (IFC); Multilateral Investment Guarantee Agency (MIFA); and International Centre for Settlement of Investment disputes (ICISD).
4. **The World Trade Organisation (WTO)**: The WTO's key objective is to help global trade flow smoothly, freely, fairly and predictably. 153 countries are members, accounting for over 97 per cent of world trade and decisions are made by a consensus of the entire membership.

    The WTO administers trade agreements, settles trade disputes, reviews national trade policies and provides developing countries with technical assistance. In an increasingly globalised world, trade flow agreements are crucial.
5. **The European Union**: The EU is a group of 27 European countries that participate in the world economy as one economic unit. With a combined population of 490 million people it has become one of the largest producers in the world, in terms of GDP, and has significant political influence and ability to drive change. Three of the global challenges are particularly important within the EU: global warming, water and population.
    - **Climate Change**: The commitment to reduce greenhouse gases by 20 per cent by 2020 acknowledges the environmental damage these cause. The EU produces about 14 per cent of the world's greenhouse emissions. A further incentive is to

ease dependence on powerful energy exporters such as Russia by a fresh wave of competition and investment in infrastructure and technology. Other binding targets are to increase energy efficiency by 20 per cent and increase the use of renewables to 20 per cent of energy consumption.

- **Water supplies**: Sustainable management of water resources is another priority. Twenty per cent of all EU surface water is seriously threatened with pollution. Sixty per cent of European cities overexploit groundwater resources, which make up the major part of their water supply. The cost of supplying the EU with water almost doubled between 1990 and 1995. Major directives have been issued on subjects such as urban waste water, drinking water, nitrates in agriculture and pollution prevention. There is a real sense that without a coordinated attack on these problems the EU will face a serious water crisis.
- **Population**: Europe is facing unprecedented demographic changes that will have a major impact on the whole of society. The EU's fertility rate fell to 1.48 in 2003, below the level needed to replace the population – 2.1 children per woman. There will be a major decline in Europe's population and the ratio of dependent young and old people will increase from 49 per cent in 2005 to 66 per cent in 2030. Modern Europe has never had economic growth without population growth. A special EU Commission is working with member states to consider all options, ranging from incentives to encourage mothers to have more babies to an acknowledgement that the gap can only be filled by increasing further the flow of migrants. These must however be considered within the global context of population management.

The EU's commitment to solving major global challenges is impressive. Through its aid programmes it provides more than 50 per cent of all development worldwide. It has agreed to increase its official development assistance to 0.56 per cent of its gross national income by 2010, on the way to achieving the UN target of 0.7 per cent by 2010. Half the additional aid will go to the most fragile countries in Africa.

Three major principles guide this aid programme:
- The reduction of poverty.
- Development aid should be based on Europe's democratic values such as human rights and democracy.
- Developing countries should mainly be responsible for their own development.

There are also a number of other organisations that play an essential role in tackling the global challenges.

6. **NGOs**: There are thousands of NGOs, which express their special interests through projects in developing countries. Some like the International Red Cross, Greenpeace, Actionaid, Amnesty International, CARE and Save the Children have global coverage.
7. **Aid programmes**: These programmes are financed and managed by individual countries, concentrating on problems and needs, which they perceive as priorities. Sometimes countries join together in aid programmes – for example, through the European Economic Community (EEC). The scale of official development assistance is significant, reaching over $100 billion in 2005[1] with the USA the largest contributor. Private contributions and charitable funds are also important.

---

1 As reported on the OEDC website: www.oecd.org/document/40/0,2340,en_2649_33721_36418344_1_1_1_1,00.html, Aid flows top USD 100 billion in 2005, OECD.

In different ways all of these organisations have an important contribution to make. Whatever criticisms are made of them should be seen in the context of their huge, creative and sustained contributions to the world's challenges. The simple test is to ask, 'What would the world be like today if these organisations in their various contributions did not exist?'

However, it is widely accepted that the outcome of many costly and well-intentioned interventions has been disappointing. NGOs have been criticised for focusing on their areas of special interest without regard for the total picture. IMF and World Bank performance is weakened by the difficulty of reconciling the fundamentally different policies and priorities of nation states.

Their professional management is often not ideal. An obvious example is that key management positions may be given to the candidate of a particular country rather than objectively choosing the best available candidate.

Increasing the strength of professional management in these organisations therefore provides an important opportunity to increase their effectiveness in addressing the global challenges.

Achieving Best Practice Global Management in International Organisations

Although there are concerns about these global players, their basic aims are sound. The challenge is to improve their management effectiveness and to concentrate on expanding and leveraging specific proven successes and best practices. The challenges are pressing and therefore all this needs to be done in a limited time span.

The following provide some observations on some of the major international players:

1. **United Nations**: The UN has been criticised for being bureaucratic and for lack of efficiency. Under pressure from the USA, which withheld its dues, major reforms have been started.
   - **Efficiency**: A special office serves as an efficiency watchdog and there is review of the large number of UN mandates which are older than five years.
   - **Balance of power**: The balance of power in the crucial Security Council is under challenge. Countries such as Germany, Japan and Brazil feel they are not properly represented in the UN. The problem is that there is little consensus among the major powers as to how international relations should be structured. In this sense the disagreements at the Security Council and its inability to intervene effectively in crises such as Darfur reflect a wider fundamental political disagreement.

   Secretary-General Ban Ki-moon has set reform of the UN as his top priority, acknowledging the new urgency required in dealing with environmental, health and security threats and challenges that respect no boundaries.

   This change is a necessary effort. However, like many of its predecessors it has limited opportunities for success. Root and branch reform requires a majority vote of two-thirds of member states including all permanent members of the Security Council which will be difficult to achieve. Apart from Chapter VII of the Constitution (which relates to the use of force to maintain peace) the states are independent and autonomous. There is no sovereignty given to the UN by any of its member states.

Therefore, although the UN has an important moral and influential role, not least through its subsidiary organisations, it does not have the authority to deal itself with the major challenges facing the planet. The practical contribution of these subsidiary UN organizations, such as UNICEF and WHO, in developing policies, researching major issues, involving many different partners in seeking solutions and giving leadership in their subject areas is highly respected.

2. **International Monetary Fund**: Until recently the IMF has been facing its own financial crisis.[2] The Fund was not generating enough income to cover its $1 billion budget. By 2010 its deficit was forecast be $400 million per annum. 15 per cent of staff was to go and there were plans to sell some its vast gold stocks.

A critical report by the Fund's own Internal Evaluation Office stated that the Fund 'is increasingly unlikely to provide financing on a sufficiently large scale to meet the demands of higher risk members'. The same report urged an overhaul of the governance structure including the recommendation that the IMF Managing Director should be open to candidates of all nationalities, not as at present a European choice.[3]

The Fund was also criticised for having 'utopian' plans, which are governed by loan conditions such as liberalisation of trade investment and privatisation of nationalised industries. There were suggestions that a 'one size fits all' approach did not take fully into account the political and cultural conditions in the country asking for help. There is too much reliance on IMF models and not enough local input. The decision-making process gives power to the wealthy countries as voting power is determined by the amount of money paid into the IMF's quota system.

However, the G20 meeting in 2009 transformed the role of the IMF. It will have its funds increased substantially from $250 billion to $750 billion and it will be allowed to issue $250 billion worth of its own quasi-currency, the Special Drawing Right (SDR). Just as important as the additional funds is acknowledgement that IMF policies and management have to improve. Larger shares of votes for major emerging countries will reflect the changing global balance of economic power. Rich nations may continue to resist any weakening of their dominant position. Non-European candidates must be given the opportunity to apply for the job of IMF Managing Director. The combination of more money and sensitive policy and management changes should make the IMF more respected by rich nations and more trusted by poor nations.

A regular independent analysis of the results of the work of the Fund is necessary to check if these new opportunities are realised in practice.

3. **World Bank**: Concerns have been raised that although negotiations with a country do take place, the reality is that the power and final decision lies with the World Bank. It is criticised for imposing its western industrialised democratic structures and values where they are not necessarily relevant.

As with the IMF, the transparencies of outcomes – exactly what happened at the front end where people are in poverty – are seen as inadequate. The Bank's Independent Evaluation Group reported in 2009 that since 1997 a third of the 220

---

2   As reported in *The Economist*, 12 April, 2008, p. 4.
3   As reported in *The Daily Telegraph*, 28 May, 2008.

projects under scrutiny had failed to achieve their goals. Moreover the goals were often misconceived, more likely to benefit middle and upper classes than those in poverty. Three-quarters of the projects for Africa did badly.

Corruption is also seen as a major issue. Working through governments in many developing countries clearly exposes the risk of corruption draining resources from their major target. Presidents of the World Bank have for many years vowed to root out the cancer of corruption but the problem still remains.

The Bank's internal auditors reported in January 2008 that over $500 million worth of Indian healthcare projects had been tainted by 'significant indicators of fraud and corruption' such as 'bid rigging, bribery and manipulated bid prices'. The Bank and the Indian Government have started investigations.[4] The Bank's independent Evaluation Group in 2009 gave the lowest possible rating for fraud-detection procedures in the $40 million aid programme called the International Development Association.

Another concern is that some of the infrastructure projects financed by the World Bank have had adverse environmental and social effects. For example, the construction of hydroelectric dams in various countries has resulted in the displacement of indigenous people.

This perhaps reflects the Bank's primary concern with economic dimensions and not the wider social impacts. The interconnectivity of the global challenges means that this is no longer viable.

In May 2008, the World Bank Growth Commission Report was published after two-years work involving 21 world leaders and experts, an 11-member working group, 300 academic experts, 13 consultations and a budget of $4 million.

The key to their findings was that it is hard to know how an economy will respond to a particular policy and the right answer in the present moment may not apply in the future. In other words, general plans for securing growth do not work – and the approach has to be country specific, even period specific. However the Report stresses that growth is a key objective since 'it can free people en masse from poverty and drudgery'. This, however, needs to be seen within the context of sustainable growth.

Once more, perspective is necessary. The successes of the World Bank are significantly more numerous than its failures and for that we should be grateful. For example, the World Bank is the largest financier in the water sector with a portfolio of $20 billion in related projects being implemented in more than 100 countries.[5]

4. **The World Trade Organisation**: An important international body which promotes free trade and stimulates economic growth. Some concerns have been raised that rich countries benefit more than poor countries. Other criticisms are that the WTO does not take labour rights, working conditions and environmental issues fruitfully into consideration.

It is, of course, extremely difficult to get agreement amongst so many different countries, each with their own priorities and agendas. However, a suggestion that final decisions should be made by a small group has proved unacceptable.

The Doha[6] round of discussions to liberalise trade in agriculture, industrial goods and services among the WTO's 135 members was unsuccessful in 2008. India and

---

4   As reported in *The Economist*, 22 March, 2008, p. 79.

5   As reported in World Bank Press Review, 4 September, 2008.

6   The DOHA DEVELOPMENT ROUND is the current trade negotiation of the World Trade Organisation (WTO). Its objective is to lower trade barriers around the world and thus allow countries to increase trade globally

China sought to retain the right to protect their farmers and manufacturers who would be vulnerable to international competition. The US – and to some extent the EU – demanded access to these markets as the price for cutting their own support for farmers. One of the dangers of the financial crisis is that free trade may give way to more protectionism. This would be a major blow to global economic prosperity.

5. **European Union**: The EU, despite setbacks, has achieved its original objective which was to ensure that there is no renewed conflict between the great European powers. With all the known limitations, the Union has enhanced the economic prosperity of its members. This would not have been possible without the willingness of member states to sacrifice some of their sovereignty for the greater good. Clearly there is now reluctance in many countries to give more of their sovereign powers to the EU.

   Even when there is full agreement on a problem, the difficulty of getting agreement on action is obvious. For example, all member states agree on the need to face up to the challenge of global warming but disagree on the costs and solutions. Ninety per cent of Poland's heating energy is generated by coal. France sees nuclear power as renewable energy while countries like Austria, Denmark and Ireland regard nuclear power with great suspicion.

   It is not easy to secure a common purpose or viewpoint on issues such as migration, energy and water within a single national state. The problem is dramatically worse when 27 countries have to agree. It is essential to establish a plan which integrates the separate issues. Such an overall plan, however, does not appear to exist at present within the EU.

   The EU aid programmes are impressive in scale and individual countries also have their own programmes. Aid grants have lower levels of conditionality than other forms of aid and grants. Aid programmes such as water usage in Africa have given useful insights to EU water problems. However, in 2002, Eveline Herfkens, then Netherlands's Minister for Development Cooperation said, 'A euro of development assistance spent via the EU is still one of the least efficiently spent.'

   In spite of substantial efforts to reform there is a large gap between policy making and implementation. There is still too much preoccupation with inputs rather than results. Shortages of skilled staff and bureaucratic delays between approval and implementation remain. The creation of the European Aid Co-Operation Office has begun to tackle these problems.

   There are critics who believe it is better for each country to manage the whole of its aid money independently rather than to try and reconcile conflicting social and political policies of so many countries.

6. **Aid programmes**: The World Bank reports that a growing number of aid organisations, funds and programmes are operating 70,000 'aid activities', each of which is in the order of $1.7 million. Each donor country hosts an average of 206 missions.[7]

   The force that drives a donor country's aid programme is essentially moral. An acknowledgement that wealthy nations have an obligation to reduce poverty and suffering in the world and to enhance human rights and freedom.

   There are other motives related to the self-interest of the donor, such as reducing conflict in areas where they have commercial interests and building 'conditionality'

---

7   As reported in World Bank Newsletter, 2 September, 2008.

into aid programmes. These require the recipient to buy goods and services from the donor. Afghanistan is a good example of these mixed motives and also illustrates the waste and frustration from lack of coordination between donors.

Another example is the US Food for Peace aid programme which is the largest food donor in the world. However, intermediaries such as shipping companies and agribusinesses must by law be American. The USA Accountability office says that two-thirds of the US food aid dollar goes to administrative overhead and just one-third buys food for the hungry.[8]

The degree of conditionality varies greatly. In 2005, Canada tied 33 per cent of its aid, Australia 28 per cent, Greece 26 per cent and Portugal 25 per cent. Neither the UK nor Ireland use this system and most northern European countries tie less than 5 per cent of their aid.[9]

Is aid effective? There are mixed opinions, including strong views that aid has caused more damage to developing countries than it has helped them, particularly in creating long-term dependency on aid. The President of Rwanda in 2009 wrote, 'Often aid has left recipient populations unstable, distracted and more dependent,' and continues, 'Don't get me wrong. We appreciate support from the outside but it should be for what we intend to achieve ourselves. No one should pretend that they care more about our nations than we do; or assume that they know what is good for us better than we do ourselves.' In her book *Dead Aid*, Dambisa Moyo has aroused worldwide debate, arguing that official development assistance has fostered dependency and perpetuated poor governance. She proposes a blend of commercial debt, microfinance, fairer trade and investment in its place. Within five years all aid funds would be cut off.

However, with all its disappointments, we believe that properly managed aid has made a significant contribution to the well-being of people in developing nations.

There are fundamental difficulties in deciding where aid should be spent. For example, the prima facie priority must be those countries where the one billion poorest people live. However, without competent governance the aid is often not spent at the 'front line' and may be 'redirected' by corrupt politicians. It is also the case that these countries are more prone to conflicts, which again makes aid programmes difficult – sometimes impossible – to implement.

The need to evaluate the effectiveness of aid is now accepted after years of preoccupation with inputs without measuring the outputs correctly.

In addition to their own aid programmes, countries contribute to multilateral funds such as the Global Fund to Fight Aids, Tuberculosis and Malaria, WHO and UNAIDS.

The opportunity exists to leverage key global business management best practices which could assist in improving aid effectiveness: A number of key questions can assist to identify areas where improvement is possible:
- **Effective business planning**: Are the aid objectives clearly defined and is the measurement of results soundly based? Many worthy programmes prove not to be sustainable. Does the aid programme strengthen institutions and improve the quality of governance? Does the programme contribute to sustainable growth,

---

8   As reported on Religion News Service, pewforum.org. 13 May 13 2008.
9   As reported in *The Wisdom of Whores*, Pisani, E., Granta Publications, 12 Addison Avenue, London, W1. 2008, p. 285.

an essential objective for long-term success? Some developing countries, most of them in Africa, have had high levels of aid dependence – in excess of 10 per cent of gross domestic product or half of government spending for decades. This has not been helpful to their long-term development. In September 2008, the Executive Director of UNICEF, Ane Veneman, said urgent action to counter bureaucracy held the key to aid delivery: 'I have met with some ministers in developing countries who tell me they spend as much as 60 per cent of their time meeting with individual donors and organisations and their staff spend more time complying with separate bureaucratic procedures than delivering results.'

- **Customer focus**: Was the programme designed in serious, full partnership with the end use recipient? This may seem obvious, but the example of East Timor is not unusual. When East Timor became independent in 2002 the US State Department and USAID each gave $1 million to be spent on an HIV programme. There were in fact only seven people diagnosed with an HIV infection.[10] The end user requirement had clearly not been researched. Is there evidence that long-term aid is removing the incentive for self-improvement and responsibility?
- **Supply chain management**: What proportion of aid is conditional, benefiting the donor when the same resources can be bought more cheaply elsewhere? Do donor countries take into full consideration the level of corruption at all levels in a potential recipient country? Failure to do so guarantees the frustration of the desired objective.
- **Paris declaration**: In 2005, over 100 ministers and heads of agencies representing donor and recipient governments and multilateral aid agencies signed the Paris Declaration on Aid.[11] This was a commitment to four key principles to improve effectiveness:
    - **Think Global: Act Local**: Do donors base their overall support on partner country's national development strategies, institutions and procedures?
    - **Effective Governance**: Do recipient countries exercise effective leadership over their development policies, strategies and coordinate development actions?
    - **Effective Change Management**: Are donors' actions more harmonised, transparent and collectively effective than they were previously?
    - **Management for Results**: Are resource management and decision making focused on results?

This is a potentially valuable framework but it requires much more detail and planned follow up. Recipient countries are cynical about promises made at large conferences. Data, a campaign group, reported that the G8 countries in 2005 promised to increase aid to Africa by an extra $22 billion to $50 billion a year by 2010. By June 2008 they had provided only $3 billion of the extra $22 billion.[12]

---

10   As reported in *The Wisdom of Whores*, Pisani, E., Granta Publications, 12 Addison Avenue, London, W1. 2008, p. 285.
11   As reported on the Wikipedia website: Wikipedia: Development Aid. 19 May, 2008.
12   As reported in the *Financial Times*, Rich lambasted over Africa aid, London, 19 June, 2008.

AID PROGRAMMES IN AFGHANISTAN

The IMF estimates that aid accounts for two-thirds of Afghanistan's gross domestic product[1] and a further $20 billion was pledged in June 2008.

However, the effectiveness of these programmes is questionable. Mr. Ashraf Ghani, formerly a World Bank executive and de facto prime minister for three years said that much of the aid since 2001 has been wasted and it had increased corruption and misgovernment. He claimed that 62 donor countries were pulling in different directions. For example, there are 16 competing international programmes to reform the justice system. Alastair McKechie, World Bank Director, said there was 'a huge issue' of the effectiveness of aid in Afghanistan and that 'little headway' had been made in the fight against corruption.[2]

The Afghan Government point out that substantial parts of pledged aid have not been delivered and that security is the main obstacle to assisting farmers. They also note that too much money is conditional and goes to donor country consultants and contractors. Aid programmes and NGOs are stripping the Government of its most capable civil servants.

Similar criticism is made of NGOs. There are over 300 now working in Afghanistan and attempts to coordinate their activities are not yet successful. Indeed, the Government, concerned about the poor image of NGOs, is currently evaluating them and legislation may follow. This should not imply that no useful work has been done. There has been a 26 per cent decline in under-five mortality through NGO-driven primary healthcare services funded by the World Bank.

These difficulties are not surprising when you consider the mixed political motives behind these aid programs, including:

- Strategic political objectives to reduce the threat of terrorism by defeating the Taliban and building a strong democratic country.
- A genuine moral commitment to alleviate the poverty created by wars and internal conflict.
- Controlling the 'poppy' crop in order to reduce global drug problems.

---

1 As reported in the *Financial Times*, Helping the Afghans help themselves, London, 12 June, 2008.
2 As reported in *The Sunday Times*, Afghans plead for $25 billion in aid as disorder grows, London, 8 June, 2008.

NGOs: The contribution of Non-Government Organisations (NGOs) is not fully appreciated and may well be more influential than state aid programmes. In 2004 the total value of NGO aid-funded activities was almost $24 billion, equivalent to over 30 per cent of overseas development aid.[13]

They are mainly funded by individuals and philanthropic organisations, whose members share their values and objectives. In some cases there are government grants. The scale of activities varies enormously, ranging from international NGOs, such as WWF, to local charities concentrated on a very specific local need.

Again, it is useful to review how global business management principles could be relevant.

---

13 As reported in *Does Foreign Aid Really Work?* Riddell, R. C., Oxford University Press, 2008, p. 259, http://www.us.oup.com/us/catalog/general/subject/Economics/Developmental/?view=usa&ci=9780199295654.

- **Effective business planning and monitoring**: The majority of NGOs do not always apply the same rigour and critical assessment that are made by other institutions. There is a tendency to focus on success and good news rather than failures, in order to keep up the motivation of their members. OXFAM and Save the Children have given leadership in producing frank and detailed performance reports and their methodologies are being shared with others. NGOs are becoming increasingly aware that transparency is crucial to protecting their integrity. A UK 'Think tank', New Philanthropy Capital, reported in 2009 that the governance of Britain's £33 billion is 'not up to scratch' and that charities could descend into crisis unless they strengthen their management boards.
- **Customer focus**: What percentage of the NGOs income goes to serve the targeted need, for example, sustainable forests, education for children in Ghana, and how much is taken by administrative costs? To what extent should NGOs spend their funds on lobbying and campaigning (for example, baby milk powder; land mines), at the expense of direct intervention on the ground? There are sometimes concerns raised that NGOs implicitly claim a moral superiority over other institutions and some question whether an NGO that is based in and substantially financed by the wealthy western world has the authority to speak for the poor and dispossessed. Keeping clear customer/end user focus provides an opportunity to address these concerns.
- **Wider partnerships**: Many international NGOs now monitor the social performance of companies and often issue reports very critical of a company. A recent trend is for companies and NGOs to work in partnership on such monitoring. How should this be managed without weakening the independence of the NGO?

Opportunities for Increasing Management Effectiveness

Whilst every organisation is different, there are some general 'best practice' global management opportunities.

EFFECTIVE BUSINESS PLANNING

- **Including Opex with Capex**: Where projects involve capital investment, for example, building road or power stations, the aid must also include long-term maintenance. Developing nations are littered with projects which started well and then ground to a halt because the operational competence and money for maintenance was not available.
- **Professionalism**: The constant challenge for major institutions such as the IMF, World Bank and UN is to manage their affairs in a thoroughly professional way, free from political pressures. This is an important concept when such organisations are so heavily politicised and therefore presents an opportunity for more effective management.
- **Fixing the 'leaky pipes'**: Dealing with corruption and high levels of administration costs is hugely important and too often brushed over. Funds for worthy projects channelled through governments known to be corrupt provide no real benefit to those in need. Large international agencies are still giving aid to national governments which are known to be corrupt and poorly governed.

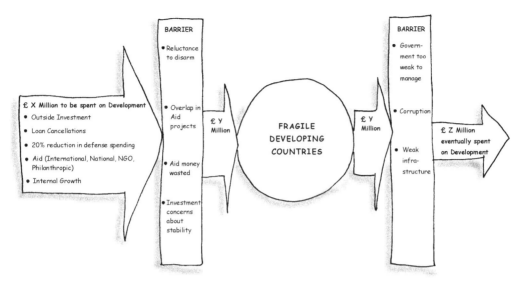

Figure 75 A problem of funding or a problem of management?

CUSTOMER FOCUS

Large plans for solving problems simply do not work if they are top down with inadequate input from those on the front line – the poor, needy and sick. So often these people do not have a voice. In business language, non-profit organisations have the opportunity to be more customer driven, focusing increased time and effort in really understanding what customers, that is, beneficiaries, need and how best it can be delivered. For example, inexpensive insecticide-impregnated mosquito nets reduce the risk of Malaria. The instinct of aid donors is to give free goods and services to those in need. However, free impregnated mosquito nets may be diverted to the black market or even used as fishing nets. An NGO, Population Services International (PSI), sells nets to mothers at antenatal clinics for fifty cents. The nurse who distributes the nets gets nine cents per net for herself, so the nets are always in stock.

Nets are sold to rich people for $5 so the programme pays for itself. In Malawi, where trials took place, 8 per cent of children under five slept under a net in 2000. By 2004 this rose to 55 per cent.[14] This success comes from understanding the realities at the front line and allowing people some dignity when they pay even a small amount for a net.

A by-product of grandiose plans is the attempt to manage every detail so that the essential actions get lost in a mass of bureaucracy. There is no evidence that aid programmes change bad governments into good governments.

80:20 PRINCIPLE/FOCUS

For every problem it is possible to identify the small number of key issues and actions which have a major influence on the successful outcome. This is where the 80:20 principle can be used most effectively. Twenty per cent of actions often bring 80 per cent

---

14 As reported in *The White Man's Burden*, Easterly, W.

success. This approach is, however, not always practised in bureaucratic environments characteristic of some non-profit organisations.

EFFECTIVE CHANGE MANAGEMENT

Good transparency is essential in effective change management and it covers everything from how key staff are selected to the proportion of funds spent on administration. Coupled with transparency is the need for independent appraisal of the organisation, its effectiveness and performance in relation to its aims. In-house departments which seek to carry out this role are a poor substitute for independent assessment. Personal accountability is fundamental with rewards given for success and failure being penalised.

A World Bank study of evaluation in 2000 admitted that, 'Despite the billions of dollars spent on assistance each year, there is still very little known about the actual impact of projects on the poor.'[15]

National Government

Although the nature of national government differs greatly between countries, their responsibility for driving constructive change is inescapable.

The opportunity to achieve this is driven by two major requirements:

- **A global commitment**: A long-term commitment to the vision, objectives and strategies is required to successfully implement a global agenda at both a national and international. The extremely limited success of the Kyoto Treaty, where a majority of countries signed up to change but real progress stalled because a minority of important countries did not, is a case in point. Cooperation at an international level requires countries to sacrifice their short-term interest for the long-term global environmental and social security of the planet. This, however, is not always politically straightforward for country leaders – which is where an understanding and communication of the global imperative is necessary,
- **National commitment to action**: A commitment at national level to understand and respond to the global challenges calls for political leadership and a sustained communication programme. It requires a good understanding of the challenges and the opportunities for action. It requires improved coordination and management performance.

Change often requires a compelling event, and the global challenges – reinforced by both their interconnectivity and the urgent timeframes in which they need to be addressed on a global scale – provide just such a compelling event. There is an unprecedented opportunity for change at both a country and international level once it is accepted that there is no other choice.

---

15   As reported in Evaluating the impact of development on poverty: A handbook for practitioners – Directions in development, Baker, J. L., World Bank, Washington DC, USA, 2000.

Each country has its own initiatives and priorities in dealing with long-term sustainability, with the western developed nations often portrayed as leaders when in fact what is required is a mutual learning process.

It is illuminating to analyse the changes which are being made in China, which as the second largest economy in the world after the USA and populated by one-fifth of the world's population, will have a major influence on the global challenges. What can we learn from them?

CHINA: CHALLENGES, PLANS AND A LONG-TERM VISION

No one questions China's rapid economic growth. In 2008, China was the world's second largest exporter and third largest importer; consumer of one-third of the world's steel production; second only to the US in car production; world's second largest importer of oil. The statistics are impressive and command respect.

However, the evidence on China's sustainability is alarming. China is the world's major carbon emitter – for example, it is the world's biggest producer and consumer of coal with coal-fired power meeting 80 per cent of energy needs. With about the same amount of water as the USA but with nearly five times the population there is already a water crisis in some cities and regions. The rivers and aquifers are draining and what is left is becoming polluted. In the World Bank's words this will have 'catastrophic consequences for future generations'. The effects of global warming and the continued growth of population add to the list of concerns.

The general impression is that these problems are not publicly acknowledged by China's political leaders because fast growth must be sustained at all costs if employment is to be maintained and living standards are to rise. However, this view is no longer correct. Plans are being made to address the sustainability challenges. Perhaps the best evidence is in an interview given to *Der Spiegel* on March 7th 2005 by Pan Yue, now number one deputy director of the Chinese State Environmental Protection Administration:

> 'Many factors are coming together here. Our raw materials are scarce, we don't have enough land and our population is constantly growing. Currently there are 1.3 billion people living in China, that's twice as many as 50 years ago. In 2020, there will be 1.5 billion in China. Cities are growing but desert areas are expanding at the same time; habitable and usable land has been halved over the past 50 years.
>
> The environment can no longer keep pace. Acid rain is falling on one-third of the Chinese territory; half the water in our seven largest rivers is completely useless, while one-fourth of our citizens do not have access to clean drinking water. One-third of the urban population is breathing polluted air, and less that 20 per cent of the trash in cities is treated and processed in an environmentally sustainable manner. Finally, five of the most polluted cities world wide are in China... Because air and water are polluted, we are losing between 8 and 15 per cent of our gross domestic product. And that doesn't include the costs for health. Then there's the human suffering. In Beijing alone, 70 to 80 per cent of all deadly cancers are related to the environment. Lung cancer has emerged as the No. 1 cause of death.'

This is an incredibly frank statement. He could have added that China's elderly population is rapidly increasing with healthcare and sanitation improvements and by 2050 about a third of the population will be aged over 60. As a result of the one child policy the working population will shrink. Nor does Pan Yue comment on the vulnerability of China to Climate Change .For example glaciers in the north west of China have decreased by 21 per cent since the 1950s.

The attack on these complex and interconnected issues was first made through the 10th Five-Year Plan. This had limited success except in one important respect: it brought sharp realisation that only by becoming sustainable could China survive and prosper is the long run. The 11th Five-Year Plan from 2006 to 2010 is more ambitious and sets specific goals such as reducing energy intensity to 20 per cent below 2005 levels by 2010.

Provincial governments have targets. There are quantified and ambitious plans to develop renewables from 7 per cent to 16 per cent by 2020. Pricing mechanisms are used with higher prices paid for cleaner fuels. It is planned to reduce the small carbon emitting power stations by 8 per cent by 2010. Provincial governments must replace 50 million traditional incandescent lamps with subsidised energy efficiency lights in one year. These are just some examples of comprehensive plans and targets for each of the global challenges.

The task is so demanding no one can be certain of its success but to put it simply: a nation which has a coordinated plan and political commitment to change has more chance to succeed than one which does not. The same can be said of the Planet.

There is also wider vision beyond seeking to balance environmental needs and the imperative of economic growth. It is by investing heavily in clean energy and other sustainability technologies China will take a leadership position in low-cost, high-performance equipment and techniques which it will export to countries which have been slow to develop them. The International Energy Agency reported in 2009 that, 'China has become the major world market for advanced coal fired plants with high specification control systems.' Such new plants are being built at the rate of one a month. It now costs a third less to build a supercritical power plant in China than to build less efficient coal-fired plants in the USA. China is already the world's largest exporter of photovoltaic solar panels and is rapidly gaining market leadership in wind energy generation equipment. In terms of national competitive advantage this is a far sighted vision given the immense new global market for green products and technologies.

CHAPTER 21
Leveraging the Business Contribution

'Our obligation as business leaders is to leave the world better than we found it.'

Sam DiPiazza, Global CEO PricewaterhouseCoopers

The growth of global corporations and the accompanying information and communications revolution has created what is now commonly referred to as the global village. Although this has sometimes led to exploitation of resources and people, the major changes have been positive. Economic growth has been viewed as an essential driver for improving living standards.

Business plays an important role in the development of this global village:

- Knowledge and best practice have been spread worldwide through capital investment and innovation.
- Business, a major user of resources, has an important part to play in their global sustainable use.
- National companies have a similar role to play within a particular country.

Small businesses have a good track record in quick, flexible response to new opportunities and innovation. They have, for example, been leaders in the development of areas such as sustainable energy resources.

Business has a critical role in addressing the ten Global Challenges. Since an estimated 70 per cent of the world's resources are used by business, it has a clear responsibility to use them in a sustainable way. Moreover, business is the most successful provider of the innovation and funding which will be required to successfully support the changes.

This section will focus mainly on private enterprise and in particular on the major MNCs. That does not reflect a lack of respect for the significant contribution being made by State Owned Enterprises (SOEs).

Opportunities for State Owned Enterprises

The scale of SOEs is substantial: the World Bank in 2006 reported that they remain important in emerging and developing countries, amounting to about 10 per cent of GDP. Definition is often difficult. In China, for example, it is estimated that there are 150 to 200 major corporations reporting directly to central government. However, they in turn have thousands of subsidiaries where the state remains the major shareholder even where the other shareholders are municipalities or private enterprises. In India there are 240 SOEs outside the finance sector. In Russia, companies controlled by the federal government produce 20 per cent of industrial output.[1]

SOEs exist for different reasons including national security and preservation of assets such as oil. It is also argued that SOEs can fill needs of little interest to the private sector. For example, keeping prices low, particularly for basic goods and services such as energy, housing, and transportation, may provide employment opportunities which are not strictly commercial. Sometimes the reason is ideological such as the belief that the capitalist system itself is fundamentally socially unjust.

However, considering the role of SOEs in contributing to the global challenges presents some problems. Many SOEs are undergoing a process of change and review. Nearly 120 countries carried out some privatisations between 1990 and 2003 alone. Serious debate is taking place in many countries as to the best approach. Is there scope for joint ventures? Where an SOE has become a major international player are the objectives primarily commercial or political? These preoccupations present some limitations in the use of time and resources for contributions to the global challenges.

The test for SOEs must always be comparison with best practice foreign competitors. In general, SOEs are two to three times less productive. This is understandable since the potential for bankruptcy is absent and subsidies are common. Political interference in setting non-commercial objectives is normal. SOEs are, however, commonly strong in utilities such as gas, water, electricity and telecommunications, and therefore have an important role to play as agents of change in government.

Particularly in developing countries, the financing needed to develop such utilities is enormous, so great is the gap between growing demand and present delivery. Attracting funds through aid or other investment channels is only the start of a complex process of reorganisation. It is unreasonable to expect such SOEs to be able to contribute substantially to global initiatives.

Acknowledging this problem, the EU has decided to play a more active role in helping partner governments in developing countries to devise and implement reform of their SOEs, especially the public utilities. OECD has set out Guidelines on Corporate Governance of State Owned Enterprises as a framework for constructive discussion.

The Business Contribution in Context

'It is the absence of broad-based business activity, not its presence that condemns much of humanity to suffering. Indeed what is utopian is the notion that poverty can be overcome without the active engagement of business.'

Kofi Annan, former Secretary – General, UN[2]

Influence formers, such as left of centre politicians, conservation organisations, charities and faith groups, are often suspicious of capitalist business and at times depict it as

---

[1] As reported in The international activities and effects of state-owned enterprises, Shapiro, D., Simon Fraser University and Globerman, S., Western Washington University, December 16, 2007. Prepared for presentation at the Centre for Trade Policy and Law Conference on 'Canada's foreign investment policies – A time for review?', Ottawa, December 6, 2007.

[2] As quoted in the International Business Leaders Forum – The Role of business in development, Davies, R. IBLF Chief Executive, Media Room Article, 28 November, 2005, www.iblf.org.

greedy, exploitive, polluting and guilty of overriding social interests. Since no institution in society is perfect it is easy to give examples ranging from Enron, the Bhopal pesticide catastrophe in India and the environmental damage from the Exxon Valdez oil spill. The late response of large pharmaceutical companies to offer low-price anti-viral drugs for HIV/AIDS sufferers is also a legitimate criticism.

Sometimes, however, these criticisms are without foundation. It was said that MNCs increase profitability by moving operations to countries where environmental regulations are weak. World Bank studies found little evidence to support this, stating that the most important factor in determining the amount of investment was the size of the local market. Within a given industry, foreign-operated plants tended to pollute less than their local peers as they are often used to more regulated environments. Business usually welcomes strong governance because bad, weak governance has damaging impacts. A World Economic Foundation (Cologny, Switzerland) Report 'Partnering to Strengthen Public Governance' (January 2008) sets out the issues clearly.

Table 12   Partnering to strengthen public governance[a]

	Governance Deficit	**Impact on Business**
Bad Governance	• Endemic social conflict • Disputed/cancelled/absent elections • High levels of corruption • Inefficient administrative and regulatory processes	• Business interruption resulting from increased volatility and conflict • Legal risks and financial outlays related to bribery and corruption • Unreasonable costs and barriers to establishing and operating businesses. • Increased cost and political risk of providing security for people and fixed investments
Weak Governance	• Lack of basic public goods supporting economic and social development (e.g. water, healthcare, education) • Weak or uneven enforcement of existing laws and regulations (e.g. labour, environment, consumer safety) • Failure to adapt rules to emerging problems and circumstances • Poorly developed conflict resolution mechanisms (i.e. impartial judiciary)	• Business interruption resulting from increased volatility and conflict • Legal risks and financial outlays related to bribery and corruption • Unreasonable costs and barriers to establishing and operating businesses • Increased cost and political risk of providing security for people and fixed investments
Underdeveloped Global Governance	• Endemic lack of consensus through existing structures (e.g. WTO Doha Round) • Absence of structures to address key global issues (e.g. Climate Change, human rights) • Conflicting or duplicative frameworks over best way to address key regional and global issues (e.g. Climate Change, natural resource management)	• Risk of diminished political support for open trading system • Heightened risk due to lack of long-term frameworks for planning capital investments • Diminished license to operate resulting from low levels of trust in business • Business uncertainty regarding rules and guidelines

[a]   As reported in World Economic Forum: Partnering to strengthen public governance, The leadership challenge for CEOs and Boards, January, 2008, p. 10. World Economic Forum, 91–93 route de la Capite, CH-1223 Cologny / Geneva, Switzerland, +41 (0)22 869 1212, contact@weforum.org, www.weforum.org, Ref: 090108.

The response of a minority of business leaders is equally strident, highlighting the historical comparison of market-driven business compared to socialised operations. They point out that dreams of self-sufficient communes have little place in the real world where globalisation is irreversible.

It is unfortunate to see resources and emotion wasted in such disagreements when there are menacing global challenges which can only be defeated by pooling experience and resources, being respectful of different contributions and seeking partnership. More humility and less criticism will be helpful moving forward. Cooperation is the key. There are no 'good guys' and 'bad guys'. This is a good example of where we are all on the same side in addressing the global challenges.

Business also needs to acknowledge the pioneering work of environmental NGOs, which have shaped world opinion and exposed threats and opportunities. The Sierra Club, with its 1.4 million members, has successfully campaigned to protect the environment in North America since 1892. WWF, launched in 1961, has 5 million members. Greenpeace, established ten years later, has 2.8 million supporters with national and regional offices in 41 countries.

How many whales would be dead today without the Greenpeace campaigns? When there is a major natural disaster such as an earthquake the first on the scene is invariably the International Red Cross. For centuries, churches, mosques, synagogues and other places of worship have been focal points of pastoral care for the poor and oppressed. They are driven to serve by love and faith whilst politicians look to political expediency and business sees no profit opportunity. An influential group, dissatisfied with the limitations of classical economics, has set up the Green Economics Institute.

NGOs have the opportunity to acknowledge the unique and essential contribution made by business. Although the focus is understandably on MNCs there are successful cooperatives, such as the retailing giant Migros in Switzerland with over 80,000 employees which plays a key role in Swiss daily life. The John Lewis Partnership is another example of a successful business partnership.

At another level there is micro-financing, where small loans are made mainly to rural poor people in the developing world to enable them to set up a small business. Professor Muhammad Yunus was awarded the Nobel Peace Prize in 2006 for his pioneering work in Bangladesh where there are now 6.6 million borrowers, of whom 97 per cent are women. Micro-financing has spread now to many developing countries.

Rotary International has made a major impact on the eradication of Polio on a global scale.

CONQUERING POLIO: ROTARY INTERNATIONAL'S PROGRAMME TO ERADICATE THE WORLD'S GREATEST CRIPPLING DISEASE

Rotary's war on polio began in 1979 with a commitment to buy and help to deliver vaccine to 6 million children in the Philippines.

In 1985 the PolioPlus programme was established in partnership with health agencies, national health ministries and philanthropic institutions. An extraordinary commitment was made to use Rotary members as foot soldiers to deliver vaccine to more than 100 million children born each year with polio in developing countries. The scale of need was enormous: the polio virus was circulating in 125 countries, paralysing or killing 1,000 people a day, mostly children.

> As the end of the twentieth century drew to a close, almost two billion children had received oral polio vaccine and only 20 countries were still polio endemic. Five million people, mainly in the developing world, who would otherwise have been paralysed, are walking today because they were immunised against polio. 500,000 cases continue to be prevented each year. In two decades the number of polio cases has been reduced by 99 per cent. The battle continues until the certification of the eradication of the virus, which will require a minimum of three years of zero polio cases following the discovery of the last case of polio.
>
> **What lessons can be learnt from this experience?**
>
> Firstly, it is a positive demonstration of how Thinking Global: Acting Local can impact a major global challenge. It can be defeated by the passion and unselfish commitment of millions of individuals working through non-governmental voluntary organisations.
>
> Secondly, it shows the power of Partnership. Although Rotary members dug deep into their pockets there was no way they could find the $500 million in donor resources to maintain the initiative, let alone develop it.
>
> However, with exceptional leadership, Rotary International brought on board many countries contributing some $1.5 billion by mid-2003. Corporate, philanthropic and individual contributions made up the balance.[1]
>
> In 2008, Rotary International, the Bill and Melinda Gates Foundation, British and German Governments committed more than $630 million in new funds with a commitment to completely eliminate the polio virus.
>
> ---
> 1 As reported in Conquering polio, Hernert A Pigman, Rotary International, One Rotary Center, 1560, Sherman Avenue. Evanston, IL 60201-3698, USA.

The United Nations Development Programme, through its Commission on the Private Sector in 2004, presented a report on 'Unleashing Entrepreneurship: making business work for the poor'.[3] Aware that small-scale indigenous enterprises are in most countries the primary engine of job creation and domestic commerce, it looked for strong expansion in sustainable private sector investment.

This would be the main driver of accelerated economic growth which is essential for reducing poverty. The co-chair of the Commission, Ernesto Zedillo, who heads the Yale University study of globalisation, stated, 'There cannot be human development without economic growth, and a fundamental ingredient of growth is the private sector.'

Success of this Development Programme will be dependent, however, on full cooperation between the public and private sector.

This framework for action for developing countries will be based on the principles of the core of the Commission report:

- Primary responsibility for achieving growth and equitable development lies with the developing countries themselves.

---
[3] As reported in United Nations Development Programme. Press Release: Unleashing entrepreneurship: Making business work for the poor, 1 March, 2004, Ph: +1 212 906 6606.

- Developing-country governments must make a strong, unambiguous policy commitment to sustainable private sector development and combine that with regulatory reforms eliminating artificial constraints to economic growth.
- Government must create real partnerships with the domestic private sector to implement the required changes and ensure that these partnerships include small and mid-sized businesses and micro-enterprises.
- There should be a clear official recognition of the 'informal' business sector.
- MNCs are formidable contributors to the world's economic growth. Definitions vary but it is estimated that MNCs account for 20 per cent of the world's industrial output and 70 per cent of world trade, from about 3,000 in 1990 there are now some 63,000 MNCs directly employing 90 million people (of whom 20 million are in the developing world).

Looking at MNCs there are some important issues to consider:

- **Regulation**: There is a misconception that these companies operate free of control around the globe. Each MNC, and its subsidiaries, work within international, national and local laws and regulations. Criticism of companies in the arms, alcohol or cigarette business is understandable but must be seen in the context that these operations can only take place with the explicit approval of governments.
- **Guidelines**: Far from regarding laws and regulations as a cost of doing business, it is MNCs which are insisting on clearer international regulations. For example, at the European Business Summit in 2008, companies in the steel, chemical and oil sectors implored the European Commission to provide more clarity on the climate regime they would face after 2013. Uncertainty was holding up investments. The Commission says it will not decide on these matters until the outcome of the international negotiations on a post-Kyoto climate regime becomes clearer.
- **Code of conduct**: There are also codes of conduct, which international business, as well as governments and trade unions, have pledged to support. The two major codes are the ILO (International Labour Organisation) Tripartite Declaration of Principles concerning Multinational Enterprises and the OECD (Organisation for Cooperation and Development).

Opportunities for Multinational Companies

'As a CEO, get on the sustainability train or you will go out of date. I think the train left a couple of years ago... The agenda has moved much faster than people realise. Tomorrow's story is what you are doing to help the planet be a better place will be part of your brand promise. Sustainability is gathering pace at the international level and companies will have a very, very big role to play.'

Mervyn Davies, CEO, Standard Chartered Bank

Opportunities cover a whole range of issues including taxation, science and technology, industrial relations, environment, health and safety. It is true the codes issued, for example by OECD, have no legal teeth but they are influential in changing behaviour. In common with business in general, MNC's unique contribution to society is that they create the economic wealth which, through taxes, enables hospitals to be built, schools to be established and infrastructures to be developed.

Some examples of best practice:

1. **The Unilever Group**: In 2006, out of a sales income of 39.6 billion euros they spent over 28.2 billion euros with suppliers, adding 11.4 billion euros added value through their operations. With 179,000 employees and 317 manufacturing sites across six continents, Unilever make a major contribution to nutrition, hygiene and personal care and community development.
2. **Innovation**: MNCs are powerhouses of technological innovation. Technologies are spreading to developing countries faster than ever in history.

    Mobile phones are an example of a new technology leapfrogging a past generation by skipping fixed-line technology. A fisherman buying small chunks of mobile phone time can check prices at the market. Small-scale electricity generation based on solar panels is another example. Large companies making local investment bring new machinery and a platform of new learning.

WHAT DOES A HOST COUNTRY EXPECT FROM AN MNC?

A host country's expectations include:

- Major MNC investments and developments should be discussed in all aspects with the appropriate public authorities to ensure they are in line with the countries development plans and priorities.
- Readiness to involve local interests and investors, public and private, in the enterprise.
- Willingness to obey the national financial and tax laws and regulations in spirit even where this is not specifically required by national law.
- The reinvestment of at least part of the profits made in the host country.
- The development of R&D/technology capability in the country at so that permanent technological dependency is avoided.
- The provision of technical licences at reasonable levels with regular reviews.
- The development of constructive positive labour policies and plans notably in providing good wages and working conditions, cooperation with employer and worker organisations, consultation and establishing plans in low employment areas.
- The willingness to assist in developing export business.

The list is indicative and will obviously vary in emphasis from country to country.

WHAT DOES AN MNC EXPECT FROM A HOST COUNTRY?

An MNC's expectations include:

- Evidence there is a genuine and long-term business opportunity.
- Political and economic stability for their investment.
- A trustworthy and clear fabric of law, rules and regulations. As the International Chamber of Commerce put it, 'Many problems and misunderstandings would disappear if

> developing nations had fully fledged laws, efficient and independent auditing systems, and fair tax systems and clear and effective labour laws.' These are far more important than special favours, tax holidays and so on which are currently offered as incentives.
> - Equitable treatment in relation to similar national companies.
> - Absence of corruption. As the UN Eminent Persons Report says, 'We are very conscious of the fact that ostensibly sound policies may be undermined by beaurocratic red tape or even rampant corruption.'
> - Some confidence that their investment will not be expropriated or at least the belief that if it is there will be a fair, prompt compensation according to well-understood rules. For example, through established international conciliation and arbitration procedures.
>
> By its nature an MNC is a risk taking institution. These and other variables will never all be right… but the balanced appraisal of risk is always made against anticipated benefits.

- **Water treatment**: Consider the contribution of a new water treatment unit which uses UV light to kill water borne bacteria, viruses and parasites – sources of the diarrhoeal disease that kill 1.5 billion children a year.
- **Food productivity**: There are calls on academic and scientific research to develop high-yield crops… but it is private business which has the resources and capability to multiply seed on a large scale and provide effective water management equipment and fertilisers.
- **R&D centres**: Where corporations set up R&D centres the impetus to spread knowledge touches local universities, technical colleges and education systems. Of course, the capacity of a country to absorb this new technology depends also on the basic infrastructure. Computers, for example, need a reliable electricity supply.
- **Management expertise**: When MNCs enter a developing country they bring new management expertise and technical and professional knowledge and skills which then cascade to local suppliers. MNCs are one of the world's most efficient agents of change.
- **Capital and investment**: MNCs also bring large sums of capital and give better access to outside financial markets and sources of credit. Trade flows often expand rapidly and provide the opportunity for developing countries to leverage their capabilities.

There has been concern that MNCs constantly move from country to country to maximise their profit, choosing countries where political power is weak. However, given the cost of establishing and dismantling company infrastructure this is usually incorrect. Of course the portfolio of investment will change but in general terms business needs to take a long-term view and looks for regulatory, political and financial stability. Start-up operations in a country call for major capital investment which is only written off when there are major strategic changes.

World Bank Managing Director, Dr. Ngozi Okonjo-Iweaala, speaking in 2008 of her experience as Nigeria's Finance Minister leading economic reform says:

'The overall goal was to promote private sector development as a vehicle for wealth creation and poverty reduction. SOEs in industries like telecommunications, petrochemicals and hotels were privatised and went from making loss to making profit. Key sectors were deregulated

and suddenly became attractive to the private sector. The results have been beneficial for the country as a whole.

The telecommunications sector for example has seen over US$1 billion a year in investments over the past four years. Nigeria went from having just 5 million telephone lines in 2001 to over 32 million GSM lines in 2007.'[4]

However, within the framework of addressing the challenges facing Planet Earth, the significant contribution business can make requires two important qualifications:

- **Business performance**: Firstly, there is great scope for business to improve its own performance since the gap between the best and the rest is far too wide. Because of their size and power, very large businesses can have a globally negative impact when they get things wrong. The flawed business models on subprime loans and apparent 'greed' of highly respected financial institutions stimulated a global credit problem in 2008 which continues to have serious human and national consequences.
- **Distribution**: Even the positive global business forces still bypass a significant proportion of the population in poor countries. Those living in fragile, broken states often weakened by internal ethnic conflicts suffer self-reinforcing cycles of poverty, poor health, limited education and lower life expectancies. Those living on less than a $1 a day remain in the order of a billion people, though the latest MDG Report indicates that this level is dropping.[5]

The leaders of global corporations recognise that their growth – indeed their long-term survival – depends on improving their performance and cooperating with other members of society such as governments and NGOs to help these underprivileged people. The opportunity has come to rethink the social and environmental contract between business and society.

BOTSWANA AND DE BEERS: A CASE STUDY

The Debswana Diamond Company, jointly owned by Botswana and de Beers, is the world's leading rough diamond producer with 27 per cent of the market. Its Jwaneng mine and the economic spin offs have transformed what was an agricultural country in the 1960s into a politically stable, substantially free from corruption, high-growth country. The former Vice President, Mr F.G. Mogae said in 1997, 'The partnership between de Beers and Botswana has been likened to a marriage. I sometimes wonder whether a better analogy might be that of Siamese twins.'[1]

In March 2008, a new jointly owned company was formed which will market and sell 15 per cent of Botswana's diamonds in polished form. This adds about 40 per cent to the value of a

---

1   As quoted on the De Beers Company website: www.debeersgroup.com.

---

4   As reported on the World Bank website: web.worldbank.org/WBSITE/.../0,,contentMDK:2 1632297~pagePK:141137~piPK:1 41127~theSitePK:226309,00.html.

5   As reported on the UNDP website: www.undp.org/mdg/basics_ontrack.shtml.

diamond. Traditionally the rough diamonds were exported to London. Fifteen international diamond firms with specialist cutting and polishing skills have been established – new outside investment. All sorting and valuing of the diamonds will be done at a new $83m state-of-the-art plant in the capital Gaborone. This is another Botswana-de Beers joint venture.[2]

Three thousand jobs will be created in a country of 1.8 million where unemployment is high. A wide range of new skills will be developed.

'What we are embarking on is nothing less than one of the largest transfer of skills and commercial activity to Africa ever seen,' said de Beers Chairman Nicky Oppenheimer.

The mine's General Manager Balisi Bonyongo says, 'We have seen a lot of opportunities being created for employment, for infrastructure, hospitals, health services – on the back of the Jwaneng Mine.'[3]

---

[2] As reported in *Time*, Mining: A gem of an idea, Perry, A., p. 57, 26 May, 2008.
[3] As reported in *The Economist*, Botswana: The southern star, 29 March, 2008.

Social Contribution: Risks and Benefits

Businesses are struggling to find the right balance. The pressures to increase contributions to society are compelling but actions must not risk damaging their unique role. By satisfying and retaining their customers, companies provide the financial means through taxes for schools, hospitals and other infrastructures. Business also makes investment into new technologies and services. Quite simply, businesses that do not operate profitably go out of existence and cannot make an ongoing social and environmental contribution. Understanding the risks and benefits and achieving the correct balance is therefore crucial.

POTENTIAL RISKS

There are a number of risks in not responding proactively.

- **Responding to the environment**: No business can be successful, indeed survive, in the long term if it is out of tune with the expectations of society. This means more understanding and action on the social and environmental constraints within which the business operates.
- **Global communication**: Social demands and criticisms are increasing in strength and volume, as individuals and lobby groups use global communications capabilities such as the Internet.
- **There is no place to hide**: Literally hundreds of independent organisations are observing and reporting to the public. At the World Economic Forum the 100 most sustainable companies are listed following systematic assessment. Oxfam has prepared a major study 'Investing for Life' giving their views on the failures of global pharmaceutical companies to respond to the needs of the poor.

- **Competitive disadvantage**: If competitors have a better track record for social responsibility this is increasingly a competitive advantage for them. A food retailer in the western world which does not offer organic products would suffer loss of business.
- **Liabilities**: Failure to live up to legal requirements and even self-imposed breaches of social responsibility policies and public statements may result in negligence claims. However, Chief Executive Officers increasingly focus on opportunities rather than risks.
- **Loss of opportunities**: Global changes give opportunities for new products and services. Significant investments are being made in wind and solar energy and pharmaceutical companies see new markets for better birth control products and treatment for HIV/AIDS and Malaria. Even when the margins are thin in developing countries, the potential market size is substantial.

POTENTIAL BENEFITS

By working in partnership with governments and NGOs, the reputation, influence and potential market of a company is enhanced. Here are a few examples:

- **General Electric**: Having already invested $3 billion in renewable energy, GE announced in 2008 a doubling of its investments in renewables. This is part of the GE 'economagination' initiative that includes pledges to increase revenues from 'green products' to $20 billion.
- **Johnson Controls**: Fifty per cent of Johnson Controls business now comes from batteries for hybrid cars and products to run buildings efficiently. Although customer interest is at present lukewarm, car makers are responding to regulatory pressures on greenhouse emissions and fuel economy. J. D. Power Associates forecast that hybrid vehicles will roughly quintuple to half a million units by 2015 under a pessimistic scenario and could near 1.2 million under optimistic forecasts.[6]
- **People matter**: Customers are increasingly aware of the social responsibility image of a company and this influences their buying decision. Staff want to work for companies that not only state socially desirable values but implement them, day by day.
- **Diageo**: Two decades ago, all the grain for Diageo's breweries in Nigeria was imported. The imports required precious foreign currency and represented a lost business opportunity for local farmers. Diageo joined a project to develop the cultivation of a beer-friendly variety of sorghum in Nigeria. The project identified a usable sorghum cultivar and trained farmers to grow the crop. Sorghum farmers reported a 35–50 per cent increase in yield from their land. Today, Diageo's breweries in Nigeria source 95 per cent of their grain from local farms, sustaining around 27,000 jobs.[7]

Companies are recognising that social and environmental responsibility presents a significant opportunity and that a robust, transparent and auditable sustainability strategy is good for business.

---

[6] As quoted in the *Financial Times*, Makers hedge bets on alternative vehicles, Reed, J., 4 March 4, 2008.
[7] As reported in Department for International Development UK. press release, 23 September, 2008.

The starting place to convert this policy into hard action is the Corporate Social Responsibility (CSR) Audit.

Corporate Social Reponsibility Audit

The case for a CSR Audit is now beyond doubt no matter what disagreements there are on methodologies. A survey of 250 business leaders worldwide conducted by the IBM Institute for Business Value in 2007 showed that over two-thirds were focusing on CSR activities to create new revenue streams. Over half believe their company's CSR activities are already giving them an advantage over their top competitors.[8]

There has been a sea change in putting CSR into practice. Literally every day brings a new pressure or initiative. There is no doubt that environmental issues, in particular the impact of global warming, have been a major impetus.

A new Companies Act was introduced into the UK in 2008 that obliges company directors to report on social and environmental risks and opportunities, including risks down their supply chains. The EU has launched a new European CSR proposal. In December 2008 a law was adopted in Denmark requiring the country's 1,100 largest companies to report on their social responsibility efforts.

It is no longer a choice. Businesses are not development agencies or charities but they are obliged to be key players in the battle to address global challenges. However this positive news must be viewed with caution.

Although the first UK book on Social Responsibility Audit by John Humble, was published more than 30 years ago, very little serious work followed. The result is that many businesses are still at a learning stage where the methodologies are not consistent or reliable. Some companies still see CSR as part of their public relations activities rather than core to the survival of their business.

However, even where there is limited knowledge about the CSR Audit there is now the opportunity to make a start, secure improvements and start learning together.

WHAT IS A CORPORATE SOCIAL RESPONSIBILITY AUDIT?

It is important not to get trapped by the fashionable label of CSR.

Some companies have a Corporate Sustainability Report. Marks & Spencer, the major UK retailer, works under the title 'Plan A, because there is no Plan B', It's what is done that counts, not the title.

The most important principle is that CSR is not an optional extra. It is an integral part of the business' vision and objectives. For clarity it is reviewed in its own right to a format suited to the business. But it is also a normal part of every other function and process of the company. This is no different from a financial audit where the finances are brought together into corporate balance sheets, cash flows and income statements but every department and function is analysed individually as well from a financial viewpoint. Similarly sales figures and customer satisfaction figures are all key components of a company's long-term viability and short-term performance.

---

8 As reported in IBM Global Business Services: Attaining sustainable growth through corporate responsibility, Pohle, G. and Hittner, J. IBM Global Services, Route 100, Somers, NY 10589.

The word audit is familiar and the guidelines that govern financial and other audits, such as quality, productivity and customer satisfaction, have the following factors in common:

- The audit covers the critical success factors.
- The audit has measures that are judged objectively.
- The audit draws on the best answers based on knowledge of those both inside and outside of the business. It calls for true partnership.
- The audit includes a search for improvements.
- The audit includes specific objectives for implementing the potential improvement with resource, time and responsibilities set out with effective feedback on progress.
- The audit report is communicated to those who need to know and is thoroughly discussed. It is accurate, balanced and clear.

Using these general guidelines, CSR or Sustainability reports provide an important opportunity for communicating with a company's stakeholders – investors, workforce and the community. The Report describes objectively how societal and environmental trends are affecting the business and in turn how its presence and operations influence the community in specific ways. These range from environmental impacts, safety, employment, water utilisation, energy utilisation and so on. It identifies strengths and weaknesses, defines achievement and lays down measurable goals for the future.

For a company that has not done a CSR audit before, there are many sources of advice from professional consultants and institutions. A good start can be made by contacting the Global Reporting Initiative (GRI) (www.info@globalreporting.org). GRI is a voluntary organisation which has pioneered since 1997 the development of the world's most widely used sustainability reporting framework.

More than 1,000 organisations, including many of the world's major brands, have voluntarily adopted the GRI Reporting guidelines. It has developed a set of core metrics of universal value together with some special features for specific industries and smaller businesses. The GRI Reporting guidelines are available free for downloading through its website.

European policymakers recognise the importance of CSR as part of their commitment to become the world's most competitive and knowledge-based economy. Best European practice was reviewed in conferences held by the EU in 2001/2 and there is now a Multi-Stakeholders Forum on CSR Portal (www.europa.eu).

CSR Europe (http://www.csreurope.org) is a leading European business network with about 75 MNCs and 25 national partner organisations as members. They are a source of valuable tools and research and provide a vehicle for the sharing of best practice.

Another valuable resource is the 'Measuring Impact Framework' available from the World Business Council for Sustainable Development website www.wbcsd.org. This framework was built 'by business for business' and reflects the collaborative work of over 25 multinational companies over a two-year period. It moves beyond compliance, encourages stakeholder engagement, is flexible, complements existing tools and has been externally reviewed by more than 15 stakeholders, including Oxfam and Harvard University.[9]

---

9 As reported in Measuring impact framework, WBCSD, 4 Chemin de Conches, CH 1231 Conches-Geneva, Switzerland.

In 2006, HRH The Prince of Wales set up an Accounting for Sustainability Project which in turn led to a major report, 'Accounting for Sustainability'. This is an invaluable resource, setting out principles and giving practical guidelines with case histories on CSR. The Connected Reporting Framework illustrates the fundamental concepts. The full report may be downloaded from www.accountingforsustainability.org.

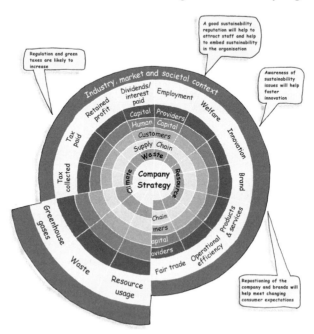

Figure 76   The Connected Reporting Framework

It is also useful to scan the reports of leading companies in relevant industry sectors and other highly regarded companies. There is often a willingness to share experience.

PricewaterhouseCoopers has a staff of 140,000 people serving its clients in 149 different countries. It has an award-winning reputation. Professionally, Pricewaterhousecoopers works with many organisations to introduce CSR systems. The firm itself in 2007 contributed $33 million to 'not for profit' and community organisations, providing the equivalent of 120 work years of voluntary service, and gave the equivalent of 83 years of free or heavily discounted professional service and community leadership.

The words of Samuel A. DiPiazza Jr., CEO, PricewaterhouseCoopers International, should be echoed by all top managers: 'As trusted advisors in the business community, we have a responsibility to consider all aspects of social and environmental sustainability... Our advice cannot be based solely on the drivers of change for today; we have the responsibility to help the drivers of the future.'

As a company learns, it develops the precise format that best suits its needs. BT's CSR Reports are world class and ranked sector leader in the Dow Jones Sustainability Index for seven consecutive years. Sixty six per cent of its staff, who serve more than 16 million customers in over 170 countries, feel proud to work for BT because of its CSR programmes. A Leadership Panel of external experts independently advises BT on sustainability and reports are assessed by Lloyds Register of Quality Assurance. The latest report can be downloaded from the BT website www.btplc.com. A summary of 2008 key performance indicators and results is shown in the following diagram.

BT Group plc: Changing World: Sustained Values 2008

Key performance indicators		Direct company impacts			
		Non-financial indicators			
		Target 2009	2006	2007	2008
Customers	Customer service A measure of success across BT's entire customer base	To improve customer service in line with the corporate scorecard and to achieve an even greater year on year improvement in Right First Time (RFT)	3% increase in customer satisfaction	3% increase in customer satisfaction	9% improvement in right first time service
Employees	Employee engagement index A measure of the success of BT's relationship with employees	BT will maintain the 2007/2008 level of employee engagement, 5.04, as measured by its annual employee attitude survey	New measure in 2007	5.07	5.07
			65%	65%	
	Diversity A measure of the diversity of the BT workforce	BT will maintain a top 10 placement of 4 in 5 major diversity benchmarks. 4 of these benchmarks will be UK based and the 5th a global diversity benchmark.	New measure in 2007	BT is in the top 10 placement in 3 out of the 4 main UK diversity benchmarks	BT is in the top 10 placement in 4 out of the 5 major diversity benchmarks
	H&S: lost time injury rate Lost time injury incidents expressed as a rate per 100,000 hours worked on a 12 month rolling average	Reduce to 0.186 cases per 100,000 hours	0.281 cases per 100,000 working hours	0.238 cases per 100,000 working hours	0.188 cases per 100,000 working hours
	H&S: sickness absence rate Percentage of calendar days lost to sickness absence expressed as a 12 months rolling average	Reduce to 2.2% calendar days lost due to sickness absence	2.35%	2.43%	2.43%
Suppliers	Supplier relationships A measure of the overall success of BT's relationship with suppliers	To achieve a rating of 80% or more next year	New measure in 2008	New measure in 2008	78%
			89%		
	Ethical trading A measure of the application of BT's supply chain human rights standard	To achieve 100% follow up within 3 months for all suppliers indentified as high or medium risk	230 risk assessments with 100% follow up	413 risk assessments with 100% follow up	213 risk assessments with 100% follow up
Improving society	Community effectiveness measure An independent evaluation of our community programme	To achieve a maximum of 82%	New measure in 2007	70%	79%
	Investment to improve society	Maintain a minimum investment of 1% of pre-tax profits	1.07%	1.05%	1.02%
Environment	Global warming CO_2 emissions A measure of BT's climate change impact (UK figures only)	To reduce CO_2 emissions to 80% below 1996 levels by 2016	0.69m tonnes 57% below 1996 level	0.68m tonnes 58% below 1996 level	0.68m tonnes 58% below 1996 level
	Waste to landfill and recycling A measure of BT's use of resources (UK figures only)	To reduce tonnage of waste sent to landfill by 6%	59,665 tonnes to landfill (58%) 42,340 tonnes recycled (42%)	54,921 tonnes to landfill (58%) 40,007 tonnes recycled (42%)	42,822 tonnes to landfill (54%) 36,937 tonnes recycled (46%)
Digital inclusion	Percentage of UK population who have not used the internet in the past three months	To reduce the percentage of people who are not using the internet to 32.4% by 31 March 2010	New measure in 2007	36%	38%
Integrity	Ethical performance measure A measure of our progress in acting with integrity	To increase to 86%	New measure in 2007	84%	85%

All targets have an end date of 31 March 2008 unless otherwise stated.

Figure 77   BT Group 2008 Values Report

TYPICAL CORPORATE SOCIAL RESPONSIBILITY AND SUSTAINABILITY ISSUES

There are often no universally accepted measurements for some of the important issues. What are the good and bad points for a local community when a major petrochemical plant is built near a small town in a developing nation? It pays to involve those who are affected to get the best answers. Increasingly, major corporations cooperate with independent and local bodies with special expertise.

There are a number of best practice examples that can be drawn on:

- **Oxfam and Unilever**: 'In 2004/5 Oxfam, an agency devoted to poverty relief, and Unilever, an Anglo Dutch consumer goods company, jointly conducted a detailed study of the economic impact of Unilever's operations in Indonesia. The conclusions were eye opening. Unilever in Indonesia supported the equivalent of 300,000 full-time jobs across its entire business, created a value of at least $630m and contributed $130 m a year in taxes to the Indonesian Government.'[10]

It is a fact of life that activists and special interest groups will publicly criticise the scope and objectivity of a CSR report. Transparency is the only response. If the accusations are legitimate then corrective action must be taken. If they are false the company must robustly defend itself publicly since its reputation is at stake. There is real danger in allowing the perception to grow that the pressure groups are always right, always standing on higher moral ground than the company. However, there is no reason to believe they are always without bias, special pleading and poor research. Environmental lobbyists forced Shell to abandon plans to scupper its Brent Spar platform in the Atlantic and instead tow it to a Norwegian fiord to be dismantled. Break up came at a high-energy cost and was a greater risk to marine pollution. The environmental groups admitted they had made a wrong judgment. There are good and bad activists and charities just as there are good and bad companies.

For a company to report on everything is wasteful .Whilst covering many issues companies can use their special experience to focus on the small number of issues where they have a major influence – 80:20 once again plays a key role.

- **Eon**: The world's largest privately owned power company focuses its efforts on cleaner energy. Its carbon intensity – the amount of CO_2 produced per unit of energy – has gone down 20 per cent since 1990 with further 10 per cent reduction planned by 2012 on 2005 figures. It is also investing heavily into renewables.
- **Staples**: Staples give high priority to environmentally friendly products and recycling. They were first in their industry to offer an online catalogue to reduce the waste of paper catalogues. They offer in-store hi-tech re-cycling every day for small businesses.
- **Zurich Financial Services**: Zurich Financial Services provides insurance products to low-income customers in Latin America. It has launched, with the Swiss Agency for Cooperation and ILO, a three-year initiative on micro-financing.
- **General Electric**: General Electric in 2005 launched a major new initiative called 'economagination' with commitment to focus on products and services which offer sustainability performance from gas turbines to light bulbs; from railway engines to jet engines. All the programmes must offer a sustainability improvement of at least 10 per cent.
- **TNT**: TNT capitalises on its logistical leadership to provide emergency response teams in a five-year partnership with the World Food Programme. They were quickly on hand to help in the 2006 Asian tsunami.
- **Alcan**: Alcan is a global aluminum producer that sees Climate Change as the top priority, not only by minimising or eliminating greenhouse gas (GHG) from its operations but by providing new technologies with wider contribution to society.

---

10   As reported in *The Economist*, A special report on corporate social responsibility – The next question: Does CSR work?, p. 8, 19 January, 2008.

Obviously each report reflects the priorities of a company and the style of presentation is very different.

- **Coca-Cola**: An overall picture is set out before getting into detailed reports. The CEO of Coca-Cola pledged that every drop of water it uses to produce beverages would be returned to the earth or compensated for through conservation and recycling programmes.[11]
- **Unilever**: The scene is set through an analysis of the impacts it makes. Detailed reporting sections follow on Unilever's performance and plans in nutrition, hygiene, integrating sustainability, wider environmental footprint, Climate Change, creating and sharing wealth.
- **Marks & Spencer**: As one of the UK's leading retailers of clothing, food and home products, with a turnover of £7.9billion, it makes a clear statement of its approach to CSR:

'Our approach to CSR underpins our commitment to Quality, Value, Service, Innovation and Trust. It allows us to manage our operations responsibly, identify risks and make the most of opportunities to differentiate ourselves from our competitors. This helps us to attract shoppers to our stores, recruit and retain the best people, form better partnerships with our suppliers and create greater value for our shareholders.'[12]

Then there are very detailed, objective statements under the following headings with regular progress reports each year. An independent assessor approves the process:

- product;
- sustainable raw materials;
- eesponsible use of technology;
- ethical trading;
- nimal welfare;
- responsible financial services;
- choice of nutritious and healthy foods;
- reducing waste from packaging and products;
- easy to understand labels and information;
- people;
- diversity and opportunity/training;
- communications and consultation;
- reward;
- health, safety and well-being;
- places;
- community;
- shopping environment;
- transport;
- reducing energy, water use and waste;
- assurance statement.

---

11 As reported in *Time*, The burden of good intentions, 23 June, 2008.
12 As reported in Marks & Spencers' CSR Report 2005, Our approach to CSR, p. 2.

CONCLUSION

This is only a brief overview of CSR and Sustainability. The positive news is that CSR/Sustainability strategies are becoming increasingly integrated into the core of progressive business – and with personal leadership from the CEO.

There can be significant opportunities for business. 'Doing well by doing good', in this case, is an opportunity to make a significant contribution to meeting the Global Challenges.

> **THE GREEN REVOLUTION IN AFRICA**
>
> The green revolution that allowed Asia to accelerate its food production in the late 1960s and early 1970s passed Africa by.
>
> Africa is the only region in the world where per capita food production has been declining for the past three decades. This is a major cause of slow economic growth since 70 per cent of total employment is in agriculture and the sector contributes 30 to 50 per cent of national incomes. Over the last 15 years the number of Africans living on a dollar a day has increased by 50 per cent.
>
> The root cause of the problem is that millions of small-scale farmers cannot grow enough food to sustain their families, let alone their communities. They lack access to high quality seeds; fertilisers for the depleted soil – Africa's soils are the poorest in the world and they lack simple water systems to deal with erratic rains.
>
> AGRA (Alliance for a Green Revolution in Africa) is a non-profit organisation, chaired by Kofi Annan, former Secretary-General of the UN, committed to breaking this cycle of hunger and poverty. Unlike many NGOs, AGRA is African-led, participatory and comprehensive. It is pro-poor and pro-environment. An early collaboration in 2006 focused on more productive and resilient varieties of Africa's major food crops such as cassava, together with education programmes for farmers. Further initiatives deal with low-cost water management systems and improvements in crop storage, finance and transport systems.
>
> An initial funding of $300 million is being strengthened by grants from national governments, the Rockefeller Foundation, the Bill and Melinda Gates Foundation and the UK Department for International Development. In June 2008, the US government Millennium Challenge Corporation (MCC), which has already invested $1.7 billion thus far in African agricultural development, joined forces with AGRA to concentrate on Ghana, Madagascar and Mali.
>
> This is a compelling example of AGRA's leadership in developing a grass roots approach to a massive challenge, in close partnership with others.

The Opportunity of the New Philanthropy

Philanthropic contributions are a long tradition for many businesses. Historically it meant allocating a percentage of the company's profits to worthy causes such as funding hospitals, local schools, orchestras and the like.

There has been a major change in approach, acknowledging that philanthropy is no longer an important 'add on' but an integral part of the company's strategic contribution as a global citizen. The best companies target funds in close partnership with their stakeholders, manage the programmes and measure the results.

- **Unilever**: In 2006, Unilever contributed 78 million euros to communities, equal to 1.6 per cent of pre-tax profit. This helped to support 13,000 community organisations around the world. In addition, Unilever has global partnerships with organisations such as UNICEF, The World Heart Foundation and the World Dental Federation, which support the company's health, nutrition and hygiene 'Vitality Strategy'.[13]
- **Eli Lilley**: Another case of a company building on one of its strengths – the treatment of TB – is Eli Lilley. A Lilly MBR-TB Partnership was established with Medecins Sans Frontieres (MSF) and WHO to provide two discounted drugs to meet immediate treatment needs worldwide. There are now operations in 40 countries and a $70 million commitment combining early detection and treatment, healthcare worker training, surveillance, community and patient support and transfer of drug-manufacturing technologies in the developing world.[14]
- **GlaxoSmithKline**: In 2009, the prices of all medicines were cut for the 50 poorest countries – the cost being less than 25 per cent of the price in wealthy nations. The company will also provide free access to its patents relating to neglected diseases.
- **Tata Steel**: When Tata Steel and other Indian companies were challenged by the then Indian Prime Minister to set aside 1 per cent of net profit for community developments, Tata Steel were able to say that over a ten-year period they had dedicated between 3 per cent and 20 per cent of profits to such work. Long term, in cooperation with local governments and other agencies, significant improvements have been made in maternity and child health, AIDS awareness, watershed management, land and water management, and enterprise promotion.[15]
- **Google**: Google funds core programmes which are fighting Climate Change, building economic development and creating an early warning system for pandemics with a $5 million grant to InSTEDD, an NGO.[16]
- **Pricewaterhousecoopers**: In 2008, Pricewaterhousecoopers donated $4 million towards the education of refugee children in eastern Chad's camps, the largest single corporate donation ever received by the UN Refugees Agency.

There have been some concerns raised that this scale of philanthropy is far from altruistic. The clear response is: does it matter if important jobs get done which would otherwise be neglected? Long-term sustainable projects need to be win-win. Within this context, however, transparency is essential.

Unilever reports that in 2006, 40 per cent of its giving went to long-term community investment partnerships. A third went in donations to charitable organisations to meet

---

13   As reported in Unilever: Sustainable Development Report 2006, www.unilever.com, Unilever PLC, PO Box 68, Unilever House, 100 Victoria Embankment, London, EC4P 4BQ, United Kingdom, Ph: +44 (0)20 7822 5252.
14   As reported on the Business Fights Aids website: www.businessfightsaids.org.
15   As reported on the TATA Website: www.tata.com: Community Initiatives.
16   As reported in *The Economist*, Face value: Google's guru of giving, p. 69, 19 January, 2008.

pressing social needs and more than a quarter of its projects were defined as seeking clear commercial benefits along with positive community impacts.[17]

PHILANTHROPIC FOUNDATIONS

Businessmen who have acquired significant wealth have created foundations which are making important contributions to social development. These foundations include:

- **The Ford Foundation**, with assets of $13.7 billion, approved $530 million in grants in 2007. The focus is on strengthening democratic values, community and economic development, education, media, arts and culture and human rights.
- **The Li Ka Shing Foundation** has a two pronged approach: capacity empowerment through education and the building of a caring society through education and medical care projects. Grants already exceed $990 million.
- **The Carnegie Corporation** of New York has education as a priority. It has assets of over $2 billion and has already given grants of some $2 billion.
- **The Bill and Melinda Gates Foundation** is the largest private foundation in the world. Founded by Bill and Melinda Gates in 2000 and doubled in size by Warren Buffet in 2006, it applies business management approaches to its giving. Its endowment is approximately $34.6 billion and each year it donates a minimum of $1.5 billion. It has three grant-making programmes: Global Health, Global Development and United States.

  To give some idea of scale: the $800 million that the foundation gives every year for global health approaches the annual budget of the UN WHO (192 countries). The sum of $287 million has been donated to 16 different HIV/AIDS research teams across the world, on the condition that they share their findings with one another. Malaria research spending has quadrupled since the Foundation gave $1 billion in 2007. In the same year, 300 of the world's Malaria experts met with the Foundation to combine their efforts to find a vaccine.
- **Cooperation**. These foundations invariably work in close partnership with governments and NGOs and also cooperate with one another. The Rockefeller Foundation has invested $10 million and the Gates Foundation $100 million to enhance agricultural science and small farm productivity in Africa.

Opportunities Presented by the Social Challenges of Supply Chain Management

Supply chain management spans all movement and storage of raw materials, work-in-progress inventory and finished goods from point of origin to point of consumption. Companies acknowledge that their success increasingly depends not only on their own social and environmental behaviour but also that of their suppliers. For this reason, understanding and effectively managing supply chain management is becoming increasingly important in establishing sustainability strategies.

---

[17] As reported in Unilever: Sustainable Development Report 2006, www.unilever.com, Unilever PLC, PO Box 68, Unilever House, 100 Victoria Embankment, London, EC4P 4BQ, United Kingdom, Ph: +44 (0)20 7822 5252.

KEY ISSUES WITH SUPPLY CHAIN MANAGEMENT

In a globalised world, the supply chain has grown to be increasingly complex, literally spanning the world through different stages involving multiple suppliers and subcontractors. Managing different cultures and expectations and controlling new risks increases vulnerability. Unilever, for example, has 8,430 suppliers of raw materials and packaging.[18] A single Chinese factory may be producing for as many as 18 different buyers.

To reduce costs, there has been an increasing move to outsource whole functions. However, even though a company may establish a contract covering social issues with the outsourcer, they may have little information about, or influence on, how the outsourcer behaves with their suppliers.

There are many activist pressure groups which seek to influence the social performance of suppliers to a company.

In the UK, a sustained campaign for Fair Trade products, that give more financial benefit to local producers in the developing world, has seen product growth of 46 per cent between 2005 and 2006 and now has 3,000 certified products. It is a multi-million pound business. The British Government has given a £1.5 million grant to the Fairtrade Foundation. The approach has its critics but it would be commercially foolish for a food retailer not to stock Fair Trade goods.[19]

The carbon footprint of suppliers can also be significant and increasingly it is becoming part of a company's social responsibility to audit this and cooperate with suppliers to reduce their company's total carbon footprint.

This is not only a part of CSR, it is also protects the company's market reputation should a campaigning NGO systematically study, audit and report on any breaches of social conduct in the company's supply chain. Apple is a good example of this. When Greenpeace identified concerns about working conditions for staff in the manufacture of iPods in China, Apple immediately reviewed its supplier's trade practices and put improvement programmes in place. Apple now has a proactive programme addressing these issues with suppliers.[20]

A jewellery business, which inadvertently buys 'conflict diamonds', can be subject to heavy criticism and customer backlash if this is identified publicly.

Companies that use timber are now aware that customers are looking for products that are certified to come from renewable timber sources.

A risk assessment should be made of the supply chain in respect to global warming as it becomes clear that major catastrophes, such as drought and flood, may have a damaging impact on supplies.

HOW COMPANIES RESPOND

Companies must rethink and restate Codes of Conduct and Conditions for each of their suppliers. These can be discussed and agreed as common ethical principles in addition to the normal legal contract. Nestle sets standards of business conduct on child labour, business ethics, hospitality, conflict of interest and sanction. It also precisely defines sustainability and environmental practices.

---

18 As reported in Unilever: Sustainable Development Report 2006, p. 23, www.unilever.com, Unilever PLC, PO Box 68, Unilever House, 100 Victoria Embankment, London, EC4P 4BQ, United Kingdom, Ph: +44 (0)20 7822 5252.
19 As reported in *The Sunday Telegraph*, Fairtrade fails to tackle third world poverty, says think tank, Gray, R., 24 February, 2008.
20 As reported on: http://management.silicon.com/careers/0,39024671,39161619,00.htm.

Even with explicit shared values and supplier contracts that embrace social and environmental commitments, it is very difficult to be sure what is happening in a long supply chain. Increasingly, audits are becoming best practice. A company itself may conduct the audits or may subcontract these to a specialist organisation which has expertise and integrity. Cooperation with charitable organisations is becoming increasingly common to ensure this transparency. Also, where a factory in a developing country produces a product for a range of different companies, these companies are cooperating to hire one independent assessor. Levi Strauss is now working with 15 other firms in 130 factories.[21] Other examples include:

- **Marks & Spencer**: Cooperated with the Carbon Trust to study the carbon footprint for the company, including suppliers. They are also studying social and economic impacts of food sourcing; setting standards for direct suppliers from farms; improving the sourcing of timber; reducing the amount of peat used for growing plants and flowers and working with the Marine Conservation Society on fish sourcing. Addressing the social impact of the supply chain fully also involves proactive action to deal with suppliers who will not cooperate.[22]
- **Unilever**: Communicated its Business Partner Code, outlining expectations on health and safety, business integrity, labour standards, consumer safety and the environment. This was supported by specific standards and audit processes. Most suppliers responded positively but in 2006 about 14 per cent did not and are now at risk to continue as future suppliers.[23]
- **Staples**: The office supply company has ended its relationship with Asia Pulp and Paper because they had no 'indication that APP had made positive strides to address serious environmental issues such as unsustainable logging in rainforests. A tough decision as APP represented almost 10 per cent of the retailer's paper stock. It demonstrates a total corporate commitment to sustainability values.'[24]
- **McDonald Corporation**: Instructs its beef suppliers 'that we do not, have not and will not permit destruction of tropical rainforests for our beef supply. We do not, have not and will not purchase beef from rainforest or recently deforested rainforest land. Any McDonald's supplier that is found to deviate from this policy – or that cannot prove to be in compliance with it – will be immediately discontinued.'[25]
- **Wal-Mart**: Some 30 per cent of its foreign purchases come from China. Although Walmart has been criticised in respect of its environmental policies, it has recently met with 1,000 of its Chinese suppliers to set environmental goals. Some 92 per cent of its environmental footprint comes from its supply chain but it only has direct control of 8 per cent of its footprint.[26]

---

21   As reported in *Time*, The burden of good intentions, 23 June, 2008.
22   As reported in Marks & Spencers' CSR Report: Plan A – Year 1 Review, p. 6, 15 January, 2008.
23   As reported in Unilever: Sustainable Development Report 2006, p. 23, www.unilever.com, Unilever PLC, PO Box 68, Unilever House, 100 Victoria Embankment, London, EC4P 4BQ, United Kingdom, Ph: +44 (0)20 7822 5252.
24   As reported in Business Respect – CSR Dispatches, No. 121, 17 February, 2008.
25   As reported in Rainforest Policy: 2006 Worldwide Corporate Responsibility Report, www.mcdonalds.com.
26   As reported in *Green Biz*, Wal-Mart to expand environmental efforts to Chinese suppliers, April, 2008, www.wbcsd.org.

CONCLUSION

Effective companies work positively to identify these issues in the supply chain as opportunities to be proactive. In practice, a rigorous critical review of the supply change from a social and environmental viewpoint often leads to reduced costs as well as the less easy to measure protection and increase in a company's reputation.

Conflict Management: Does Business Help or Hinder?

Violent conflict within countries devastates the lives and human rights of millions of people, destroys social and economic infrastructures and increases poverty. National government and the international community have the prime responsibility for addressing this problem but there is growing recognition that business has an important part to play. This is particularly true of MNCs. Fighting corruption, financing social projects such as schools and health facilities for local communities and ensuring wise investment decisions, particularly those which affect basic services such as water, power supplies and communications, are all examples of a company's influence in a country.

The constructive role of business is inevitably weakened by maverick companies, which flourish on or at the edge of law. They deal in arms, narcotics, stolen goods and cooperate with dishonest governments to exploit the population at large. The need for better regulation and control lies with national and international politicians. Effective businesses welcome the stability such regulations provide. Many major companies are already deeply involved in regions of conflict: International Alert reported that in 2000:

- There were 72 countries where the security risk for the majority of locations in which foreign business operates is rated medium, high or extreme.[27]
- Multinational companies were investing more than $150 billion in nearly 50 countries that fall below the intermediate point in Transparency International's Corruption Perception Index. In other words, countries that may be confidently described as fairly to very corrupt.[28]

In considering how to behave in such conditions, the OECD 'Guidelines for engaging Companies in Conflict resolution' has had great influence.[29] The UN has also made specific recommendations on conflict and business, as have many western governments, such as the US and UK. The following practical advice for companies involved in conflict areas has been provided by International Alert in their Report 'Transnational Corporations in Conflict Prone Zones' written by Banfield, Haufler and Lilley in September 2003.

---

27 As reported in International Alert: Council on Economic Priorities: The business of peace – The private sector as a partner in conflict prevention and resolution. International Alert, 346 Clapham Road, London, SW9 9AP, www.international-alert.org, Ph: +44 (0)20 7626 6800.

28 As above.

29 As reported in International Alert: Promoting a conflict prevention approach to OECD companies and partnering with local business. OECD DAC Conflict, Peace and Development Co-operation Network Briefing Paper, March, 2004, International Alert, 346 Clapham Road, London, SW9 9AP, www.international-alert.org, Ph: +44 (0)20 7626 6800.

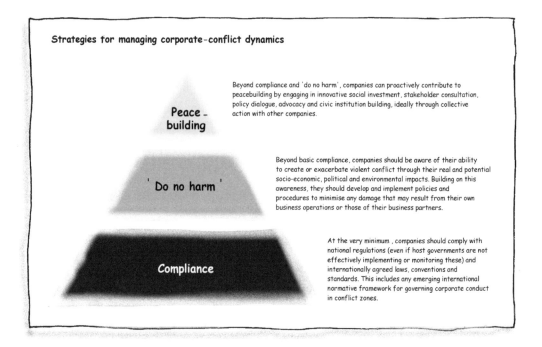

Figure 78 Strategic Management for Managing Corporate Conflict Dynamics
Source: International Alert: Transnational Corporations in Conflict Prone Zones: Public policy responses and a framework for action, p. 24. September 2003. Jessica Banfield, Virginia Haufler, Damian Lilly.

Kofi Annan, when he was Secretary General UN, in 2001, said:

'I encourage member states and the private sector to support the Global Compact in the context of the UN conflict resolution prevention efforts. In particular I encourage the business community to adopt socially responsible practices that foster a climate of peace in conflict prone-societies, help prevent and mitigate crisis situations and contribute to reconstruction and reconciliation.'[30]

Corruption: A Global Challenge for Business

Corruption also has significant impacts on business – from its reputation to its bottom line. The World Bank has highlighted that bribery is already a trillion dollar business.

- It is estimated that developing countries lose on average $20–40 billion each year through corruption and money laundering.
- Every $100 million of stolen assets returned to a developing country could fund immunisation for 4 million children, water connections for 250,000 households and Malaria treatment for 50-100 million people.[31]

---

30 As quoted in above, p. 43.
31 As reported in News and Broadcast: World Bank Managing Director calls for collective action against corruption, 6 February, 2008, http://web.worldbank.org.

Framework for types of business engagement

CORE BUSINESS ACTIVITIES
1. Implement social and environmental policies and management systems which include guidelines on human rights, anti-corruption and use of security forces
2. Undertake pre-investment conflict impact assessments and monitor real impacts on an on-going basis
3. Consult with stakeholders on a systematic basis
4. Ensure diversity in workplace practices and hire local people
5. Aim for widespread wealth creation and support for local livelihood opportunities

SOCIAL INVESTMENT
1. Build capacity of local civil society organisations
2. Invest in community-based development and participation
3. Support local education, health and enterprise development programmes
4. Fund activities that promote diversity, tolerance and civic education

POLICY DIALOGUE
1. Engage in dialogue with other companies and governments to address national development needs and tackle structural issues that underpin conflict such as corruption, inequality and human rights abuses
2. Fund think-tanks and research on these issues
3. Undertake publicity and media campaigns to promote peace

What can business do specifically?

1 Prevention strategies

Status of country or situation — STABLE PEACE / CONFLICT PRONE / CONFLICT RIDDEN

2 Post-conflict reconstruction and reconciliation

What can business do specifically?

CORE BUSINESS ACTIVITIES
1. Provide commercial support in rebuilding infrastructure and investing in productive sectors
2. Do so in a way that builds local human capital and business capacity, especially for smallscale businesses, and respect diversity

SOCIAL INVESTMENT
1. Focus on projects that target affected populations and ex-combatants, taking into account points 1-4 at top of page
2. Support NGOs active in reconciliation efforts, voter education, etc.

POLICY DIALOGUE
1. Participate in truth and reconciliation commissions
2. Support weapons hand-in, amnesty and demobilisation programmes
3. Provide funding and managerial support to build the capacity of government services, including judicial systems and police forces
4. Support initiatives to attract foreign investment to post-conflict regions
5. Build capacities and governance systems for the local private sector

3 Crisis management

What can business do specifically?

CORE BUSINESS ACTIVITIES
1. Supply relief products, equipment and services on a commercial basis in areas such as:
 • Water and sanitation
 • Shelter and site planning
 • Food, nutrition and health services
 • In doing so, follow Red Cross guidelines for humanitarian assistance
2. Ensure integrity of the company's own security arrangements when operating in a conflict zone

SOCIAL INVESTMENT
1. Partner with NGOs and governments on product donations
2. Support work of humanitarian and development efforts in other ways

POLICY DIALOGUE (Collective action)
1. Put pressure on politicians to negotiate or to remain out of regional conflicts (other than as peacekeepers)
2. Provide secretarial services and logistical support for peace negotiations
3. Engage directly in peace delegations or negotiations if appropriate and within an agreed framework

Transnational Corporations in Conflict Prone Zones: PUBLIC POLICY RESPONSES AND A FRAMEWORK FOR ACTION

Figure 79 Framework for types of business engagement
Source: International Alert: Transnational Corporations in Conflict Prone Zones: Public policy responses and a framework for action, p. 25. September 2003. Jessica Banfield, Virginia Haufler, Damian Lilly.

- The African Union estimates that the continent loses as much as $148 billion a year to corruption. Nigeria alone, in 1999, was losing $300 million each year from corrupt practices in public procurement. This money is rarely invested in Africa but finds its way into the international banking system to be laundered. The true figures globally are unknown but are of extraordinary levels.[32]

Kofi Annan, former Secretary General UN, stated: 'Corruption hurts the poor disproportionably by diverting funds intended for development, undermining a government's ability to provide basic services, feeding inequality and injustice and discouraging foreign investment and aid.'[33] In India, for example Centre for Media Studies (CMS) report says that a large section of households had to pay bribes to use public services in 2005 for police, land administration, judiciary, electricity and government hospitals.[34]

However, the impacts and damaging effects are even wider. Corruption weakens the rule of law, diverts resources from productive use, distorts local and national economies, makes start-up costs unnecessarily heavy and often causes social tensions. It has clear links with organised crime and global money laundering. Corruption is a key element in drug smuggling which feeds the drug addiction of 5 per cent of the world's population. It generates a culture where integrity is not valued. In the definition of Transparency International bribery is 'the abuse of entrusted power for private gain'.[35]

Who is responsible? The easy, justified targets are corrupt governments, particularly in developing countries and officials who are often badly paid. However, it takes two to tango! There is an opportunity for businesses to live up to their rhetoric and have in place strong policies and controls over corruption. However, even in large corporations, such policies are not always respected. Some of the most respected MNCs have been embroiled in corruption scandals. Siemens, for example, is facing damaging accusations and investigations about large sums spent in bribery around the world. The OECD anti-bribery group sent a letter in June 2008 to the British Government, attacking its failure to bring a single overseas bribery case to court, or to deliver on a years-old pledge to update its anti-corruption laws.[36] The US firm Halliburton and Co in 2009 paid a $599 million fine to end an investigation of its former KBR Inc. unit where accusations of kick-backs to Nigerian officials were involved.

Bribery, with appropriate 'labelling', can be state bribery with flagrant cases ignored because of wider political concerns. Western banks have participated in laundering the corrupt stolen funds. In some countries corruption is literally a way of life: a daily experience to get a car licence or to move home.

The Paul Volcker Enquiry Report in 2005 found that 2,253 companies, many of them household names, had made illegal payments totalling $1.8 billion to the Saddam regime, during the UN's oil-for-food programme.[37]

---

32   As reported in African Focus Bulletin, UK/Africa: The damage we do, 13 July, 2005, www.africa.focus.org.
33   Statement on the adoption by the General Assembly of the United Nations Covenant against Corruption. As quoted on: www.unodc.org/unodc/en/corruption/index.html.
34   As reported in Tracking corruption in India 2005, CMS Saket Community Centre, New Delhi, India.
35   As reported in Transparency International: Business Against Corruption – a framework for action, p. 7, www.unglobalcompact.org. www.ibill.org. www.transparency.org.
36   As reported in *Financial Times*, OECD hits out at lack of action on corruption, London, 18 September, 2008.
37   As reported in *The Economist*, The UN's oil-for-food scandal: Rolling up the culprits, pp. 82–83, 15 March, 2008.

Two UN officials have been charged, as have many businessmen. However, only 28 of the 52 countries named have chosen to review the incriminating evidence. What does this say about the determination of the international community to drive out corruption? The failure to address corruption in an open manner and manage it well reflects not only moral failure of all those involved but is also a high-risk strategy in an increasingly transparent world.

There is, however, some welcome if long overdue progress in addressing the risks and challenges for businesses in this area.

- **United Nations**: Building on earlier initiatives, such as the OECD convention on Combating Bribery of Foreign Public Officials in International Transactions; the African Union convention on preventing and combating corruption, The Council of Europe Criminal Law Convention on Corruption and The Inter-American Convention against Corruption, the UN added a tenth principle to its UN Global Compact stating:

  'Business should work against corruption in all its forms, including corruption extortion and bribery.'[38]

  There followed in 2005 the UN Convention against Corruption (UNAC) which is the first global anti-corruption instrument that addresses criminal law enforcement, international legal cooperation, asset recovery and monitoring. Each signatory must establish effective anti-corruption practices. Under the Convention, signatories agree to cooperate with one another in every aspect of the fight against corruption.

  This has become the world standard for anti-corruption efforts: a resource and clearinghouse for information and a catalyst to action.

  By January 2008, 140 countries had signed the convention and 107 of those have ratified it. This has created a valuable momentum of action in the countries concerned. However, it is early days and the gap between agreeing to the Convention and starting anti-corruption initiatives will not easily be filled considering the long-established culture and special interests involved. The challenge now is to agree proper monitoring systems, performance measures and transparency.
- **Impact on the bottom line**: It is in the self-interest of business to take pro-active anti-corruption measures. A company's values statements quickly become liabilities if these are not translated into daily action. The cost implications clearly vary between companies but Transparency International say, 'There is clear evidence that in many countries corruption adds upwards of 10 per cent to the cost of doing business and that corruption adds as much as 25 per cent to public procurement.'[39]
- **Global legal liabilities**: Legal risks, if bribery is discovered, may apply not just to the home country concerned, but have wider repercussions: the home country may well face legal sanctions against the use of corrupt practices in another country. The US Foreign and Corrupt Practices Act of 1977 established the principle and it has legal standing in the OECD and some other countries.

---

38   As reported in Transparency International: Business Against Corruption – a framework for action, p. 7, www.unglobalcompact.org. www.ibill.org. www.transparency.org.

39   As reported in Transparency International: Business Against Corruption – a framework for action, p. 5, www.unglobalcompact.org. www.ibill.org. www.transparency.org.

- **Corporate reputation**: If corruption is exposed, the company's reputation is damaged and this may prove to be a competitive disadvantage. Transparency International in their report 'Business against Corruption' proposes an outline action plan.

Transparency International Six-Step Implementation Process

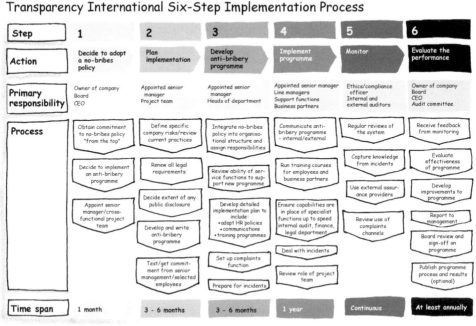

Figure 80 Corruption: Transparency International Six Step Process
Source: Transparency International: Business Against Corruption – a framework for action, p. 11. www.unglobalcompact.org. www.ibill.org. www.transparency.org.

- **Partnerships**: Another major development was launched in 2004 by the World Economic Forum through its Partnering against Corruption Initiative (PACI). Its mission is to develop multi-industry principles and practices, which will result in a competitive, level playing field, based on integrity, fairness and ethical standards. Its achievements are already substantial with 131 signatories.

The destructive impact of global corruption cannot properly be addressed without the full involvement of international, national and local governments, non-profit organisations and business. Business has the opportunity to provide leadership particularly now as the guidelines and standards are becoming clear and the price of failure increasingly high.

Investing to Fight the Ten Global Challenges

To fight the challenges of Climate Change, sustainability, energy, stabilising the population, eradicating poverty, health for all and access to education demands huge financial investment – and the efficient use of these investment funds.

National governments, through their policies on taxation, social priorities, subsidies and the like have the major responsibility for providing the capital for the essential investments. As reviewed earlier, this provides the opportunity to drive changes in the developed world and a commitment to reduce conflicts – thus refocusing resources to address the challenges.

Although these challenges are clearly limited by political decision makers, business also has an important role and there are encouraging developments.

- **Investment in sustainability**: The UN Environment Programme (UNEP) reported in 2008 that investment in environmentally friendly projects was rising.[40] Executive Director, Achim Steiner, said, 'Hundreds of millions of dollars are now flowing into renewable and clean energy technologies and trillions more are waiting in the wings, looking to governments for a new and decisive climate regime – post 2012.'[41] The Kyoto Protocol's clean development mechanism (CDM) aimed at encouraging the transfer of clean technology to poor countries has more than 850 projects, worth just over one billion dollars. Schemes worth a further 1.4 billion dollars are in the pipeline.[42] The scale of need is indicated by the United Nations Emergency Redistribution Programme (UNERP) view that, 'Investment in research and development in low-emissions technology has to triple to 90 billion dollars per year by 2015 and to 160 billion dollars by 2025, in order to stabilise C02 concentrations at 550 parts per million.'[43]

  Given clear guidelines the capital can flow from the private sector as commercial opportunities are developed. Political indecision is the major problem right now.
- **World Bank partnerships**: Another example is the way in which the World Bank is forging links with socially responsible investors. In 2007 the World Bank partnered with two European banks to issue bonds designed to appeal to private investors who want to support green technology, environmentally friendly development and anti-poverty programmes, and earn a good return on their investment. ABN-AMRO, for example, sold about 150 million euros of World Bank Eco 3 Plus bonds within a month of their launch.[44] Business investment cannot replace government investment but it is an indispensable partner.

Expanding Partnerships

'Shared goals, from market building to environmental protection and social exclusion, have enabled us to build unprecedented partnerships among business, governments, civil society, labour and the United Nations…. We are partners in the understanding that in our globalised world, many challenges are too inter-connected and complex for one sector to face alone.'

UN Secretary – General Ban Ki-moon

---

40 As quoted on the World Business Council for Sustainable Development website, UNEP calls for end to barriers on fast-growing 'green economy', AFP, 20 February, 2008. www.wbcsd.org.
41 As above.
42 As above.
43 As above.
44 As reported on The World Bank website, World Bank forges links with socially responsible investors, http://go.worldbank.org.

There has been a dramatic, far reaching and large-scale increase in business partnerships with community groups, NGOs, labour and academia.

Conflicting objectives and suspicions can occur without shared knowledge and experience, mutual respect and a willingness to listen and act for the common good. The global challenges of hunger, climate warming, poverty and the like can not be beaten by one partner alone. There has been significant progress and a number of best practice examples.

- **Carbon risk management**: Three of the largest US banks created a carbon risk management protocol with The Carbon Trust.
- **Sony and WWF**: Sony Corporation jointly hosted a meeting of business leaders with WWF, the environment group, to urge governments to take more leadership in fighting global warming.
- **Logistics and WFP**: Logistics firms, Agility, TNT and USP, shared their experience and set up a logistics Emergency Team in association with the World Food Programme. These teams will intervene to help with warehousing, transportation and airport coordination when natural disasters such as earthquakes occur.
- **Avon and UNDFW**: Avon Products is working with the UN Development Fund for Women, and making a $1 million contribution to advance women's empowerment globally.
- **Michelin and local communities in Brazil**: At the end of 2001, Michelin faced serious problems of low productivity and decreasing price of natural rubber in its hevea tree plantation in Bahia, Brazil. A commitment to stay in the area required investment in a sustainable agricultural programme. This involved creating 12 family-owned plantations and donating land for a new village. The original 600 employees are still working and there are an additional 150 new jobs. Natural rubber production has increased by 11 per cent. Michelin worked in partnership with the new owners, local officials, NGOs, unions, banks and public authorities to achieve these results.
- **Mobile Telephone Networks (MTN) in Nigeria**: Through its foundation, in three years MTN has launched over 100 community and health projects, in cooperation with 26 partners in 28 of the 36 states in Nigeria. These projects have helped over a million people.
- **African Health Force Partnership**: In 2005, The World Bank's African Health Workforce received $1.6 million from the Bill and Melinda Gates Foundation and the Government of Norway to work in partnership with African countries to deal with the shortages, poor distribution and low productivity of health workers. In addition, there have been some large and highly influential global initiatives.
- **The World Economic Forum (WEF)**: This is an independent international organisation, committed to improving the state of the world by engaging leaders from the public and private sectors and wider civil society in partnerships to shape the global, regional and industry agendas. It has been a major catalyst for change with over a thousand member companies and global initiatives in health, education, chronic hunger, Climate Change and corruption. The WEF has had significant influence on policy at international and national levels and stimulated innumerable hands on practical improvements throughout the world.

- **The UN Global Compact:** This compact brings together UN agencies, labour, civil society and governments to advance ten universal principles. The explicit mission is to realise a core, sustainable and inclusive global economy through responsible business practices. The ten principles are summarised in Figure 75. It now has over 6,000 stakeholders including over 5,000 businesses, in 130 countries. It is the world's largest voluntary corporate citizenship initiative.

    The Global Compact seeks to combine the best properties of the UN, such as moral authority and convening power with the private sector's market-driven, solution-finding strengths.

    These general aspirations have been implemented in multiple, on-the-ground, practical changes as well as bringing key players together to set the agenda on fighting corruption, human rights, Climate Change, sustainable development, operating business in conflict zones and responsible investment. For example, the Global Compact Principles for Responsible Management Education has been supported by 200 business schools. As Ban Ki-moon, Secretary General UN put it, 'As teachers, you can ensure that tomorrow's leaders understand that long-term growth of a business is tied to its environmental and social impact. As scholars, you can produce research that drives innovation and helps management to recognise the benefits of being a responsible business.'

    Membership standards are rigorous. Some 800 companies have been delisted.

The United Nations Global Compact asks companies to embrace, support and enact, within their sphere of influence, a set of core principles in the areas of human rights, labour standards, the environment and anti-corruption. The principles are as follows:

Human Rights

Principle 1	Businesses should support and respect the protection of internationally proclaimed human rights; and
Principle 2	make sure that they are not complicit in human rights abuses.

Labour

Principle 3	Businesses should uphold the freedom of association and the effective recognition of the right to collective bargaining;
Principle 4	the elimination of all forms of forced and compulsory labour;
Principle 5	the effective abolition of child labour; and
Principle 6	the elimination of discrimination in respect of employment and occupation.

Environment

Principle 7	Businesses are asked to support a precautionary approach to environmental challenges;
Principle 8	undertake initiatives to promote greater environmental responsibility; and
Principle 9	encourage the development and diffusion of environmentally friendly technologies.

Anti-corruption

Principle 10	Businesses should work against corruption in all its forms, including extortion and bribery.

Figure 81   The Ten Principles of the UN Global Compact
Source: United National Global Compact. www.unglobalcompact.org.

Progress has been significant, and as one senior manager said to me, 'It is literally a miracle. Organisations and institutions that wouldn't even speak to each other a few years ago are now in productive partnership.'

Business in the 21st Century: Corporate Global Citizens

'We need business to give practical meaning and reach to the values and principles that connect culture and people everywhere.'

Ban Ki-moon – Secretary General United Nations

Business operates in a turbulent, fast-changing world and it influences, and is in turn affected directly and indirectly by, the key challenges facing Planet Earth. The growth of globalisation, driven by technology transfer, free trade, communication and transport revolutions, and now green issues, is inexorable. The activities of the human race are more interdependent and interconnected than they have ever been. Whilst MNCs lead the way, medium-sized companies increasingly source and market internationally.

However, in this fast-changing environment, the international community, in spite of major initiatives through the UN and other institutions, is struggling to find common policies and commitments. National self-interest so often inhibits sensible policy, which requires an urgent global response. Kyoto is a good example. The scale of military expenditure is another. The collapse of the DOHA round of trade negotiations is yet another

A by-product of political shifts in power is the increasing growth in influence of business, particularly the MNCs. Business operations are inextricably woven into the fabric of the solutions to the key global challenges. Even the most vocal critics admit that private business can be a key player in creating growth and sustainable development. Business creates the pool of wealth for social programmes, is a source of innovation and new technology, and creates employment.

However, increased global influence and transparency trigger demands for higher standards of behaviour and governance. A vast array of NGOs, charities and government bodies scrutinise business and insist on greater transparency and accountability.

What is slowly, and as yet imperfectly, developing is the opportunity for a new social contract between business and society. The fundamental contribution of creating wealth for society remains unchallenged.

To leverage this, business has to develop better governance and not just work to the letter of international and local law and regulations. The spirit of the new social contract is the willingness to share experience, to listen and to genuinely cooperate.

This cooperation leads to increased respect between traditional stakeholders – investors, employees, customers and suppliers – and builds stronger partnerships with government, community and society generally. This is particularly important when the market-driven system is based on trust, honest sharing of experience and mutual respect.

At this stage of the learning curve it is not surprising that some businesses are concerned that onerous wider social commitments may hinder their capacity to create profit, the ultimate source of wealth for the community. They are concerned that business will be criticised for failures that are the responsibility of governments. However, success in addressing the global challenges will be dependent on partnerships across all of these groups and, within this, business has a key role to play.

Professor Michael Porter, of the Harvard Business School and Mark Kramer, Managing Director of FSG Social Impact Advisers provide a useful framework:

> 'Corporations are not responsible for all the world's problems, nor do they have the resources to solve them all. Each company can identify the particular set of societal problems that it is best equipped to help resolve and from which it can gain the greatest competitive advantage.'[45]

When a well-run business applies its vast resources, expertise and management talent to problems that it understands and in which it has a stake, it can have a greater impact on social good than any other institutions or philanthropic organisation.

Businesses in the twenty-first century now, more than ever before, have the opportunity to operate as true Global Corporate Citizens. Never has this been so important, as we seek to build a sustainable world in the limited timeframes available.

---

45 As quoted in *Harvard Business Review Spotlight* – Making a Real Difference, Strategy and Society: The link between competitive advantage and corporate social responsibility, Porter, M. E. and Kramer, M. R. www.hbr.org.

CHAPTER 22
Leveraging the Power of the People

'Never doubt that a small group of thoughtful, concerned citizens can change the world. Indeed it is the only thing that ever has.'

Margaret Mead, Anthropologist

People facilitate change on the planet in at least four important ways – as consumers, investors, voters, and as global corporate citizens.

- **Consumers**: Demand for sustainable products and services from consumers will drive business to develop and provide them, and will also provide the incentive for innovation required to develop these services. Companies which are not sensitive to the changing needs of their customers in the long run go out of business.
- **Investors**: Large and small investors are important drivers of business strategies and innovation. The development of Green Investment Funds is a response to a greater awareness of environmental problems. Some pension funds will not invest in tobacco or arms companies.
- **Voters and lobby groups**: Political leaders survive by being sensitive to the views of the electorate and lobby groups in democratic countries. This 'people power' can directly influence government policies.
- **Global corporate citizens**: The emergence of the global village means that many people see themselves as global corporate citizens. This may be expressed through aid donations to assist those in poverty in developing countries and through voting for parties which have a global, not just a national, agenda.

Building Bridges: Working Together

Success will only be achieved by leveraging the dynamic relationships and shared responsibilities of government, business and people as illustrated in the Triangle of Change.

It is necessary to have an overall plan, broken down into actions for each of the key players in the Triangle of Change and then applied at international, national, community and individual levels.

The challenge moving forward is to take these best practice approaches and apply them on a global scale. This, however, provides the opportunity for unprecedented cooperation as we work together to build a sustainable world.

Figure 82   We've all got the solution!

EVERY JOURNEY STARTS WITH A FIRST STEP

80 million people have so far been affected by HIV/AIDS worldwide and the disease is running out of control in the poorest countries. Faced with such a daunting challenge, individuals who want to help often feel powerless to do anything. This is a simple example which demonstrates how individuals in small organisations and groups can partner with other organisations to leverage their spiritual power and practical capabilities to make a real difference.

In the UK, two churches, the combined Parish of St James, are a flourishing Christian community with a congregation of over 700 people. As an expression of their responsibility to address the global challenges, St. James, in 2005, established a Love Africa Project. This was a plan to be of practical help to HIV /AIDS sufferers in Uganda. Highlights of the project:

- Customer focus: A full analysis of customer needs based on visits to Uganda.
- Partnership: The group partnered with two Christian NGOs already established in Uganda: ACET (AIDS Care Education and Training) which had already given support to more than 5,000 AIDS sufferers and educated more than 10,000 healthcare professionals and 7,000 teachers; and Tearfund, committed to transforming lives and overcoming poverty of 50 million people in ten years, through a network of 100,000 churches.
- Funding: Sharing the vision with the congregation resulted in raising over £200k, of which £100k has already been used.
- Focus: Carefully selected projects, for example, each with precise objectives and follow-up visits and so on.
- Business planning: A long-term plan, not a one-off contribution.

What lesson can be learnt from this project which is typical of many spiritually-driven contributions by faith groups?

- Partnership: with the local people, between the NGOs and St. James and partnership with the whole congregation.

- Funds and support go straight to the end users, regardless of faith group. There is no waste which so often drains the strength of aid programmes.
- Professional management: Use of the best practice global management principles with a clear vision, objectives, strategies, financial and information plans and controls established.

Work like this is an inspiration, demonstrating that every person can join with others to play their part in addressing the global challenges. Every journey starts with a first step.

CHAPTER 23
Embracing the New Green Revolution

> *'We have everything we need to save the earth with the possible exception of political will.'*
>
> Al Gore

The actions the human race takes over the next 20 years will be critical in ensuring that the global challenges are successfully managed; this presents the current generation has a unique and awesome responsibility.

The positive news is that there is now sufficient understanding of the nature of these challenges, their interconnectivity and an increasing recognition that there are many solutions and best practices to make rapid progress. As we have seen, there are many actions taking place around the globe, however, if we are honest about the situation they are not being well coordinated. Our real challenge, as we have seen, is now to improve our organisation and coordination, to ensure that the actions already taking place, as well those that are being developed and deployed, will achieve the rapid change required in the limited timeframes available. However, everybody can embrace this New Green Revolution and contribute to its success: to be part of the new wave of innovation and global transformation, and in the process build a sustainable world, not only for today's generation, but also for our children and grandchildren. Bjorn Stigson, President of the World Business Council for Sustainable Development – the world's leading business organisation focused on business and sustainable development – states, 'What we have now is a rare chance to reshape our world. I believe we are living in a moment of history, that all round us is a new industrial revolution is beginning... It will be clean because we know we cannot go on polluting as we have been and maintain functioning ecosystems.'

This presents not only a challenge for every individual, but also an opportunity to stand up and be counted. These revolutionary waves of change are illustrated in Figure 81 which is reproduced with permission from Nicholas Brealey Publishing from the book by Steve Tappin and Andrew Cave, *The Secrets of CEOs* (2008).

286 Developing a Plan for the Planet

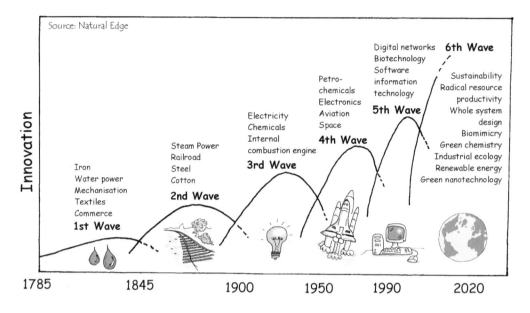

Figure 83   **The waves of change that have revolutionalised human civilisation**

These great waves of change are truly revolutionary, radically changing society at large and the way we work. They are often disturbing in the short term. As Mao Tse-tung said, 'Revolution is not a dinner party, not a essay nor a painting, nor a piece of embroidery; it cannot be advanced softly, gradually, carefully, considerately, respectfully, politely, plainly and modestly.' If a society does not have the will to change and constantly reinvent itself it will decay.

CHAPTER 24: Embracing the Spiritual Imperative

Faith Driven Power

> *'The new millennium you have just entered will decide the fate of the human species. The first two decades of this century are likely to be decisive. How that we face the prospect of extinction as a species during this period but we will set, irrevocably, the direction that will determine the survival or the demise of human life as we know it.*
>
> *Surely the divine source of all life, which many people on Planet Earth call God, could not have presented you with a more paradoxical challenge.'*
> Maurice Strong, Secretary General of the UN Conference on Environment and Development (2000)

> *'See my good works, how fine and excellent they are! All that I created, I created for you. Reflect on this, and do not corrupt or desolate my world; for if you do, there will be no one to repair it after you.'*
> Midrash Ecclesiastes Rabbath 7/13

The challenges facing Planet Earth, from global warming to poverty, from health to the environment, can be viewed through many different windows such as political, technological, social and cost-benefit.

However, the spiritual imperative is often overlooked. This is surprising since what each person believes – really and whole-heartedly believes – governs their behaviour and how they view the world. The world's faiths have great diversity in their beliefs but have in common two central tenets:

- The human race is a steward of the earth not its owner. Each person is responsible for sustaining and nourishing the earth as their legacy for future generations. It follows that selfish, short-term exploitation of the planet's resources that causes irrevocable damage is morally wrong.
- 'Love they neighbour' is expressed in different ways but every faith highlights that each human being has a responsibility to others. Every decision about our way of life and how we interact with nature has to be tested against this belief. In spiritual terms, the fact that two-thirds of the world's population live in poverty is not just an issue of social justice. It is a breach of that commitment to care for one another.

Consider the expressed beliefs on environment of the four major world's faiths: Christianity (2.1 billion members), Islam (1.5 billion members), Hinduism (900 million members) and Buddhism (360 million members).

Buddhism

The Buddhist path is one of choosing the middle way, where you consume only what you really need. Where you choose products mindfully, and with consideration of their impact on other human beings, sentient beings and the planet. Buddhism also teaches dependent origination – that every thought, every action has a consequence. What is done today will create your global future.[1]

Christianity

Christianity teaches that all of creation is the loving action of God, who not only willed the creation but also continues to care for all aspects of existence. The Report of the World Council of Churches General Assembly (1991) states:

> 'The divine presence of the Spirit in creation binds us human beings together with all created life. We are accountable before God in and to the community of life, an accountability which has been imagined in different ways: as stewards and trustees, tillers and keepers, as priests of creation, as nurturers, as co creators. This requires attitudes of compassion and humility, respect and reverence.'

Hinduism

The sacred text the Bhagavad-Gita states that a life without a contribution towards the preservation of ecology is a life of sin and a life without specific purpose or use. Dr. Sheshagiri Rao highlights:

> 'The natural harmony that should exist in the play of energies between humanity and the natural world is now disrupted by the weakest player in the game – humanity.'

Islam

Nuhammad Ihsan Mahasneh, on behalf of the Muslim World League, writes that taken in totality the principles of the Koran text states that Allah, the One True God, is the Universal God and the Creator of the Universe and, indeed, the Owner of the Universe. To Him belongs:

---

[1] As reported on Buddha101.com.

'All the animate and inanimate objects, all of whom should or do submit themselves to Him. Allah, in His wisdom, appointed us, the creatures that he has conferred with the faculty of reason and with free will, to be His regents on earth.'[2]

The Practical Power To Influence Opinion

To illustrate what this means in potential scale and influence, take the example of the Catholic Church which has about 1 billion members worldwide.

The Catholic Church has rapidly moved Climate Change, biotechnology, trade justice, pollution and the like to the forefront of its social teachings.

Pope Benedict XVI, speaking at a Vatican Conference on Global Change in 2007, attended by scientists, environment ministers and leaders of various religions from 20 countries, spoke of the need for sustainable development and warned that 'disregard for the environment always harms human existence and vice versa'.

This clear statement and commitment is then expressed worldwide through a vast network of churches with one and a half million priests and monks who are already trusted leaders. Add to that their network of schools, social centres and the like.

This example is paralleled by the other faiths. For example, The Sikh scripture declares that the purpose of human beings is to achieve a blissful state and to be in harmony with the earth and all creation. In practice, the 13 million-strong Sikh community in India maintain 28,000 temples which provide free meals to about 30 million people a day. They challenge the division of society by caste. In a territory where land resources are limited they have strong traditional values of sharing water and forest resources.

The fate of Planet Earth is now high on the agenda of all faith groups and this concern is being expressed in public statements, conferences, lobbying, and representation at key negotiations.

This commitment is becoming a major influence on public and political opinion. For example, 86 Evangelical leaders in the USA have called for federal legislation to reduce carbon dioxide emissions. The Coalition on the Environment and Jewish Life has been active in promoting community education on Climate Change.

Equally important is direct involvement of religious leaders in international, national and local institutions and on-the-ground projects. Places of worship are seeking to reduce energy use and setting an example to their members in undertaking a more sustainable life style.

Faith without action is dead.

Working Together

For centuries, faith groups have in various ways been involved in the major issues of health, poverty, and environment.

James D. Wolfensohn, then President of the World Bank, wrote that the 11 faiths which make up the Alliance of Religions and Conservation (ARC) are:

---

[2] As reported the ARCWORLD website: http://www.arcworld.org.

'Serious stakeholders in development. They are also the oldest institutions in the world and possess wisdom about how to live and how to keep hope alive.'

Graeme Wheeler, the World Bank's Managing Director for human development, says it is obvious that faith-based groups are making a tremendous difference to peoples lives. He notes that, 'In many African countries, you provide 30 to 70 per cent of the health services. And in post conflict counties, the majority of primary education services. You deliver innovation and results, and you speak courageously about ethical challenges in justice and failures in governance. You have enormous influence on family and individuals beliefs, attitudes and perceptions.'

Historic differences between science and religion, a residual sense that the spiritual world must to some extent distance itself from the material world, and theological differences within and between religions have limited the full use of faith potential to meet the planet's challenges.

This has been changing at a rapid pace. For example, following an initiative by HRH Prince Philip, International President of the WWF in 1995, 11 faiths and the World Bank now work together on a wide range of ecological and development issues in the Alliance of Religions and Conservation (ARC).

A few examples of projects given special recognition by WWF and ARC are:

- Hindus restore sacred forests in Orissa, India.
- Buddhists reintroduce traditional hunting and logging bans in Mongolia.
- Christian United Methodists have committed $40 billion to ethical investments.

Of course there are differences of opinions and priorities but the focus is common concerns and the working slogan is:

'Come, proud of what you bring of your own, but humble enough to listen.'

The power to influence public opinion is shown by the fact that the 11 ARC faiths represent two-thirds of the world's population. They own around 7 per cent of the habitable surface of the planet, they have a role in 54 per cent of all schools and their institutional share of the investment market is in the range of 6–8 per cent.

So when considering 'Who is going to drive through the necessary changes to save Planet Earth?' it is important to acknowledge that a great powerhouse of concern, based on deep spiritual values and beliefs, is beginning to be felt.

Moreover, throughout human history, churches, temples, synagogues and other places of worship have been at the front end of delivering help to the sick, oppressed and needy. The 'on-the-ground' network of each religion provides literally millions of focal points where there is respected leadership, local language and knowledge and the deep understanding of needs which only come from shared suffering. These centres will be there long after the aid programmes end and the NGOs have gone home.

Along with the direct work of the faith organizations, every individual with a sense of moral responsibility can be motivated to express his or her beliefs as a citizen, consumer, voter, politician, business person and activist.

CHAPTER 25
Delivering Our Plan for the Planet

'Unless someone like you cares an awful lot, nothing is going to get better. It's not.'
The Lorax, Dr Seuss

'Never doubt that a small group of thoughtful, committed citizens can change the world; indeed, it's the only thing that ever has.'
Margaret Mead

'If you want to go fast go alone. If you want to go far, go together.'
An African proverb

Now is the time for action. We have seen how we can effectively develop a vision for a sustainable world, translate this into a set of objectives to address the challenges we face, as well as build strategies and action plans to make it happen. We have also seen how we can translate these into specific plans for each of the key players in making this happen – business, government and people. However, unless we convert this vision and these plans into effective action, as Joel Barker states, 'Vision without action is just a dream'.

Building a sustainable world can therefore no longer be looked at as 'just a dream' – it is now an essential requirement in defining our future. It is the defining challenge of our generation. What sort of world will we leave to the future inhabitants of our planet? What did each of us do to build this sustainable world? This will be the question each of us must answer to our children, and their children. Failure to successfully address the challenges we face – as a civilisation, as a human race, as a planet – is simply not an option.

Perhaps our greatest challenge is delivering our Plan for the Planet. For as we have seen, the challenges we are facing are now compelling. Our timeframes are immediate. What we do in the next one to two decades on the key global challenges –population growth, climate change, energy, water, food, biodiversity, health, education, conflict management and financing sustainability – not only determine the type of world we will live in, but also the world we leave to future generations.

Therefore we need to move quickly and effectively to translate our Plan for the Planet into real action at a local, as well as globally coordinated scale. Yes we will need to improve, refine, develop, localise, and customise our plan – this is the nature of any plan – for no plan is perfect. However, whilst we do this, we must also be taking action. Though some may ask the question, 'Can we really do this on a global as well as local scale?', as Martin Luther has expressed so well, history is littered with 'the bleached bones and jumbled residues of numerous civilisations' over which are written the pathetic words: 'Too late'. Our key question is therefore not 'Can we do this?' Our real question is 'How can we do this?' How can we deliver our Plan for the Planet efficiently and effectively? We now know we have to deliver. We have no other choice. We know we have to be successful.

However, the good news is, we know that we can be successful in developing and delivering global plans. As highlighted, success with examples such as the Marshall Plan following the Second World War, and our global initiatives on the Global Financial Crisis, mean that we know we can deal with global challenges. We know how to build a better world – if we work together to a common plan which we all sign up to.

The exciting thing therefore about addressing these global challenges is that although no one of us can solve these issues individually, by each of us playing our part, we can can solve them together. Each of us can play our part – and has a part to play. There are no bystanders in addressing these challenges. If we choose to do nothing, we are continuing to contribute to the problems.

However, if we each take leadership and responsibility for the areas we can impact, whether this is in our business, organisation, community, household or government – we can play our part in contributing to the solutions. We can therefore all be part of the global movement to build a sustainable world. The defining task of our generation is something we can all be a part of. Now that is exciting!

So what are the key steps we need to take to deliver our Plan for the Planet? What can we do individually, and together to achieve this. Fortunately, the approach is not complex, and is something everybody can take part in.

This approach can be summarised in what we call the 'Five Ps' for delivering our Plan for the Planet.

Figure 84   Delivering our Plan for the Planet

Personalise: The first step in delivering our Plan for the Planet is to personalise it. We need to move from this being a Plan for the Planet to Our Plan for the Planet. We each need to personalise the challenges, the plan and the solutions. Work out the areas you can get involved in, what you can do to make a difference in these areas. Then work out where you can make this difference – in your business or organisation, in government, in your community, in your household. Work to understand the issues and solutions that work for you. Look at where others are being successful and how you can translate this success into success in your situation.

We no longer have the luxury of assuming that it is 'somebody else's' responsibility to address the challenges. Personalising therefore means 'owning' the areas where you can make a difference. We need to work together as one team to build a sustainable world, and as we have seen, successful teams have team members who each take responsibility for the areas where they need to, and make an impact. We need to personalise the challenge of our generation. We each need to own the part we can, and need to play a part in rebuilding a sustainable world. Putting into practice the actions you can take to build a sustainable world needs to become your personal challenge.

Prioritise: Having personalised our Plan for the Planet, it is then key that you prioritise. Prioritisation is important in two key areas. The first of these is prioritisation of time. Put time aside for your personalisation of your plan. Time to understand the issues, develop solutions and identify where you can make a difference, and monitor how you are performing. This applies equally to your business or organisation – building sustainability into your business agenda, or your household agenda. We all know if we don't put time aside for something, it probably won't happen. However, if we do, change can begin to take place at a surprising pace.

The second is to prioritise the areas where you can make a contribution to – which can have maximum impact. We each cannot do everything, yet we each can do something. Using the 80:20 rule, and the Plan for the Planet Prioritisation Matrix you can quickly determine the areas which you believe have maximum importance, maximum urgency, and where you can have maximum impact – and focus your efforts on these. To assist with this, our Plan for the Planet Prioritisation Matrix has been included in this book (see overleaf), and for your convenience, can also be downloaded from the Plan-for-the-Planet Website.

Plan for the Planet Prioritisation Matrix

Each individual and each organisation can start by identifying the global challenges that you can influence, make a difference in, and have maximum impact. These can be prioritised in terms of both urgency and importance on a scale of 1 to 10. Once this is completed, simply plot the answers on the following table to establish the areas in which to focus efforts. You can then use the table to complete a simple action plan - focusing on the strategies and actions that you can begin to take: how, who and by when. A simple exercise, but the beginning of the journey into understanding how everyone can contribute to developing and implementing a Plan for the Planet on a global scale.

No.	Challenge	Importance	Urgency	Total	Priority	Comments
1	Stabilising Population					
2	Climate Change					
3	Sustainable Energy					
4	Sustainable Water and Energy					
5	Sustainable Biodiversity					
6	Poverty					
7	Global Health					
8	Global Education					
9	Global Peace					
10	Financing Sustainability					

Figure 85   Plan for the Planet Prioritisation Matrix

Prioritisation can also be enjoyable; can build employee and organisational loyalty and family fun, as we work together towards an objective we can all believe in and share – building our sustainable world.

Plan: Once you have prioritised the areas where you can make a contribution to, and where you can take action in, it is important to then build these actions into your own personalised Plan for the Planet. Good intentions are often lost, as we know, when not translated into an action plan. Why? We all know the reasons. We get distracted, other things take our attention, and before we know it, nothing has been achieved. The benefit of putting together a personalised action plan is that it captures not only the good intentions, but more importantly, translates these into:

- your specific objectives in each of the areas;
- your strategies to achieve them, and action plans;
- your timelines for making them happen;
- who needs to be responsible for taking these actions;
- who you need to be working with;
- measurements so you can tell how well you are progressing against your objectives.

To assist with this, the Developing your own Plan for the Planet templates have also been included in the book.

DEVELOPING YOUR PLAN FOR THE PLANET (GOVERNMENT, BUSINESS OR PERSONAL)

Objectives and Strategies

No.	Objectives (What am I aiming to achieve?)	Strategies (How am I going to achieve this?)	Actions (What do I have to do to achieve this?)	Outcomes (Results)	Cost ($$)	Time-frames	Responsi-bilities	Status
1.0	**Population** • A peak population of 8.0 billion people by 2050							
	• A total sustainable world population at an appropriate living standard through natural causes							
2.0	**Climate Change** • Limit temperature increase to 2°							
	• Greenhouse gas reduction of 80% by 2050							
	• Return to preindustrial levels long term							
3.0	**Energy Supplies** • Transition to long-term sustainable and non-carbon based energy sources							
	• Maintain the 80% reduction objective in greenhouse gas emissions							
4.0	**Water and Food Supplies** • Establish long-term water security at national and international levels							
	• To halve the proportion of people who suffer from hunger by 2015							
	• Establish long-term sustainable food supplies							

Figure 86   Developing your own Plan for the Planet

Again, feel free to copy these, or download them from the website, and complete these in the areas you have prioritised as being the key ones where you can make an impact. Also, remember developing your plan can be more enjoyable, and is usually much more effective, if developed with others. These can be collegues, your business unit, company, family members and community organisations.

Participate: This brings us to the next of our key steps in delivering your Plan for the Planet – Participation. When you have completed your personalised Plan for the Planet, it is then key that you, and those you need to work with in delivering your part of our Plan for the Planet, get active in making it happen.

Participation is about working together. You are therefore encouraged to share your plan with others and to encourage them to also understand the challenges, and to understand that we call can, and need to, get active in addressing them. Encourage others to develop their own personalised Plan for the Planet. We can then not only share ideas – we can share success and through this build confidence that we can overcome these challenges together. We don't have to feel overwhelmed, we can feel empowered.

Taking an active part in building a sustainable world is therefore something each of us – our households, our communities and our countries – can take an active part in. We don't need to be bystanders – we can all be part if it.

This is the opportunity of a lifetime – the opportunity of our generation. We are being offered the very real chance to change the world, to make the world a better place, by changing the things we can change. By taking this approach, by joining our Plan for the Planet every one of us can be actively making a difference in building a sustainable world.

Figure 87 Everyone can play their part in delivering our Plan for the Planet

The real value in taking the Plan for the Planet approach is we can begin to understand that we can all participate. This approach allows us to know where to start, where we can put our focus, where we can make a difference. Tackling these challenges doesn't have to be a daunting task, it becomes an exciting opportunity to all get active in the challenge of our generation – to build a better world.

The other advantage of working together on a common Plan for the Planet is that it also helps us coordinate these efforts for maximum impact. Successfully delivering any plan depends on breaking down the overall plan for everyone into managable parts that can then be coordinated on a global scale. This makes this approach easy to use, easy to implement and easy to coordinate. Ease of use is one of the key factors in the successful implementation of any plan.

For this reason, the Plan for the Planet website has been set up so you can share your plans, your solutions and your successes. In this way, as we share success, we can build confidence, and the excitement that together we can work towards our common goal.

Let's get Passionate: This leads us to the last of our 'Five Ps' for delivering our Plan for the Planet – Passion. Great changes on our planet have not only been achieved with great plans, and great participation, they also have been driven by passion.

This is a chance to take part in one of the most exciting journeys we have made as a human race, as a civilisation – the journey towards a sustainable world. We will achieve this not only with a good plan, a plan that we all take an active part in making happen, but a plan that we execute and deliver with passion. With a good plan for the planet, good participation and passion to make it happen – we can change the world.

That is the challenge for each of us, and for our generation. This is the challenge against which each of us, and our generation, will be held accountable for by those who come after us. Did we as individuals, as a generation, stand up and take leadership in addressing the challenge?

The leadership, the action and the delivery rest with each of us. Yet as we each begin to act individually, this in turn, as we have seen, will generate the momentum, the drive and the passion to inspire others and to continue to inspire ourselves.

We have seen the success of many mass movements – we are faced now with the most important one – a global passion for building a sustainable world…

…because in the end, the most important reason each of us must passionately take up the challenge and the opportunity of acting now…

Figure 88   One of the many reasons we should be working together on our Plan for the Planet

Source: The Kids, Chambers, Ian.

…is that our generation, every one of us, has the responsibility of determining the future we leave to our children and all future generations, of not only the human race, but all living creatures we share with this treasure we call Planet Earth. Their future is in our hands…

…because Planet Earth is the only home we've got!

Figure 89 **Planet Earth – It's the only home we've got!**
Source: NASA.

'Now this is not the end. It is not even the beginning of the end. But it is, perhaps, the end of the beginning.'

Winston Churchill

A CHILD'S CHALLENGE

Hello, I am Severn Suzuki, speaking for ECO – the Environmental Children's Organisation. We are a group of twelve and thirteen year-olds trying to make a difference: Vanessa Suttie, Morgan Geisler, Michelle Quigg and me. We raised all the money to come here five thousand miles to tell you adults you must change your ways. Coming up here today, I have no hidden agenda. I am fighting for my future. Losing my future is not like losing an election or a few points on the stock market. I am here to speak for all generations to come. I am here to speak on behalf of the staving children around the world whose cries go unheard.

I am here to speak for the countless animals dying across this planet because they have nowhere left to go. I am afraid to go out in the sun now because of the holes in the ozone. I am afraid to breathe the air because I don't know what chemicals are in it. I used to go fishing in Vancouver, my home, with my dad until just a few years ago when we found the fish full of cancers. And now we hear of animals and plants going extinct every day – vanishing forever. In my life, I have dreamt of seeing the great herds of wild animals, jungles and rain forests full of birds and butterflies, but now I wonder if they will even exist for my children to see. Did you have to worry about these things when you were my age? All this is happening before our eyes and yet we act as if we have all the time we want and all the solutions. I'm only a child and I don't have all the solutions, but I want you to realise, neither do you… You don't know how to bring the salmon back up a dead stream. You don't know how to bring back an animal now extinct. And you can't bring back the forests that once grew where there is now desert. If you don't know how to fix it, please stop breaking it!

At school, even in kindergarten, you teach us how to behave in the world. You teach us: not to fight with others, to work things out, to respect others, to clean up our mess, not to hurt other creatures, to share – not to be greedy. Then why do you go out and do the things you tell us not to do? Do not forget why you're attending these conferences, who you're doing this for – we are your own children. You are deciding what kind of world we are growing up in. Parents should be able to comfort their children by saying 'everything's going to be all right,' 'it's not the end of the world,' and 'we're doing the best we can.' But I don't think you can say that to us anymore. Are we even on your list of priorities?

My dad always says, 'You are what you do, not what you say.' Well, what you do makes me cry at night. You grown-ups say you love us, but I challenge you. Please make your actions reflect your words. Thank you.

Source: Speech by Severn Suzuki, aged 12, at the 1992 Earth Summit.

JUST LET ME BREATHE

A hundred million years of sunshine
A hundred million years of rain
Can never take away the agony
Can never take away the pain
Of a hundred years destruction
The rape of mother earth
For another days production
Destroy the land that gave us birth
Just let me breathe, just let me breathe
Sixty thousand fires are burning
On any single day
There's smoke on the horizon
Ten thousand miles away
The forest is dispensable
So the politicians say
Lay bear the indefensible
And take our breath away
Just let me breathe, just let me breathe[1]

Ian Chambers (1996)

---

[1] Chambers, I., Ride the Tiger Album by Altered States.

Epilogue

> *The global economic downturn has exposed the extent to which markets and societies are increasingly interconnected and interdependent. We the participants of the B4E (Business for the Environment), recognise that the economic, environmental and social challenges and risks we face demand a new level of leadership and cooperation.*
>
> *We are confident that by exercising such leadership, restoring trust and by working together we have the opportunity to put our global economy, our markets and lifestyles and security, and, ultimately our planet on a sustainable path.*
>
> The Green Imperative, B4E Summit, Paris (2009)

The financial crisis and economic downturn have dominated the thoughts and action of leaders of all international and national governments and businesses

At first sight this is negative news for the focus and investment in dealing urgently with the global challenges.

Global setbacks… and future opportunities in the age of green economics.

Governments facing pressures to maintain the living standards of citizens and aware of the political risks of increasing taxes may well delay or reduce their promised payments to help those in poverty. Businesses fighting for their survival may reduce their commitments to greenhouse emissions and downgrade their efforts in corporate social responsibility.

Our view is that governments will temporarily fail to fully honour their pledges at a time when crucial decisions which influence the survival of life on this planet in the long term have to be made now. When the immediate panic and crises reduce, as they will, the realisation that these challenges are not an optional extra but are of fundamental importance will return.

For example, there is no escaping the enormous economic challenge of transforming the world's energy source to zero carbon systems and reducing gashouse emissions. The Obama administration is dedicating over $150 billion in green infrastructure investments over the next ten years. Short-term problems can not disguise the huge future opportunities. Ban Ki-moon, UN Secretary General, puts this in historic perspective: 'We have experienced great economic transformations throughout history: the industrial revolution, the technology revolution, and the era of globalisation. We are now on the threshold of another – the age of green economics.'

The Challenge of the World's Poor

World Bank President, Robert Zoellick, insists that global finance leaders 'must look beyond the financial rescue to the human rescue… poor countries are already facing a triple hit as the financial challenges continue to spread and many are at a tipping point... the poorest can not be asked to pay the biggest price.'

In 2009, the emergency arising from the financial crisis could force an additional 55 million people in the developing world into poverty – on less than $1.25 a day. The number of chronically hungry is expected to climb to over a billion, reversing the recent gains in fighting malnutrition. Net private capital flows to developing countries are already in sharp decline. Remittances are down about 8 per cent.

Therefore it is imperative for wealthy nations to respond decisively to this crisis, as well as dealing with their own issues.

Business Commitment to Sustainability

The good news is that progressive major businesses will not reduce their commitment to sustainability. They understand that failure to respond to the challenges which surround sustainability will injure their long-term profitability. Moreover, they see increasing opportunities as the unstoppable move to the green revolution creates the need for new products and services and forces a rethink of present practices. These often produce savings and increased efficiencies. Corporate social responsibility and all it stands for is now deeply embedded into the core strategies of many successful businesses.

The driving force for continued investment and commitments may be more economic than moral: but it will be there.

In good times and bad times the sustainability issue has to be well managed. As Georg Kell, Executive Director of the UN Global Compact puts it, 'Restoring confidence and trust in markets will require a shift to long-term sustainability. And corporate responsibility must be an instrument towards this end.' This comment increasingly reflects mainstream business thinking.

'Sustainability will remain critical to our business even during an economic downturn,' says Ian Cheshire, Group Chief Executive at Kingfisher. 'As a major international retailer we have a responsibility to tackle issues such as Climate Change and work towards a more sustainable future.'

Jan Babiak, Global Climate Change and Sustainability Leader at Ernst and Young says that companies should realise that there are substantial savings to be made from cutting down on the wasteful use of energy and economising with other materials.

Francis Sullivan, Deputy Head of Corporate Sustainability at HSBC says, 'Managing risk and promoting business opportunity is critical and, in the context of sustainability issues, is as important today as it was before the current challenges in the financial markets surfaced.'

Charities and NGOs

What of the millions of individuals who regularly and generously contribute to the work of charities and other organizations and who are researching and working in a vast range of areas related to sustainability? As personal incomes are reduced by the impact of recession there is already some fallout in spite of the fact that donations are a small proportion of personal income. This funding problem coincides with a greater demand for help from the charities. Their response must be to improve their management effectiveness and focus their reduced capability on the key issues which give maximum leverage and benefit.

The Opportunity

A potential criticism of a Plan for the Planet is that it is too idealistic. To have any hope of being achieved it requires a massive, global, coordinated and shared strategy and action plan. Although it is urgent and failure to act will damage the lives of everyone on earth, such a concept and approach is a great idea but is it 'an impossible dream'?

The global response to the worst economic and financial crisis since the Great Depression has demonstrated what is possible when world leaders acknowledge a common threat. What has resulted has been unprecedented national and international coordination, transparent information and rapid responses to the urgency.

This is encouraging proof that when a global threat is understood and acknowledged by the world's government, business and community leaders, urgent and constructive actions can be quickly put in place and globally coordinated.

The short and long-term threats of the combined ten global challenges are clearly greater than the financial crisis alone.

We have the opportunity to work urgently together to continue to develop and implement this Plan for the Planet to build a sustainable world for this generation, and all of those to come.

Appendix 1: Millenium Development Goals

About the MDGs: Basics

What are the Millennium Development Goals?

The Millennium Development Goals (MDGs) are eight goals to be achieved by 2015 that respond to the world's main development challenges. The MDGs are drawn from the actions and targets contained in the Millennium Declaration that was adopted by 189 nations and signed by 147 heads of state and governments during the UN Millennium Summit in September 2000.

The eight MDGs break down into 21 quantifiable targets that are measured by 60 indicators.

- Goal 1: Eradicate extreme poverty and hunger
- Goal 2: Achieve universal primary education
- Goal 3: Promote gender equality and empower women
- Goal 4: Reduce child mortality
- Goal 5: Improve maternal health
- Goal 6: Combat HIV/AIDS, malaria and other diseases
- Goal 7: Ensure environmental sustainability
- Goal 8: Develop a Global Partnership for Development

Goal 1: Eradicate extreme poverty and hunger

Indicators

Target 1a: Reduce by half the proportion of people living on less than a dollar a day

1.1   Proportion of population below $1 (PPP) per day
1.2   Poverty gap ratio
1.3   Share of poorest quintile in national consumption

Target 1b: Achieve full and productive employment and decent work for all, including women and young people

1.4   Growth rate of GDP per person employed
1.5   Employment-to-population ratio
1.6   Proportion of employed people living below $1 (PPP) per day
1.7   Proportion of own-account and contributing family workers in total employment

Target 1c: Reduce by half the proportion of people who suffer from hunger

1.8   Prevalence of underweight children under-five years of age
1.9   Proportion of population below minimum level of dietary energy consumption

Goal 2: Achieve universal primary education
Indicators

Target 2a: Ensure that all boys and girls compete a full course of primary schooling

2.1 Net enrolment ratio in primary education
2.2 Proportion of pupils starting grade 1 who reach last grade of primary
2.3 Literacy rate of 15–24 year-olds, women and men

Goal 3: Promote gender equality and empower women
Indicators

Target 3a: Eliminate gender disparity in primary and secondary education preferably by 2005, and at all levels by 2015

3.1 Ratios of girls to boys in primary, secondary and tertiary education
3.2 Share of women in wage employment in the non-agricultural sector
3.3 Proportion of seats held by women in national parliament

Goal 4: Reduce child mortality
Indicators

Target 4a: Reduce by two thirds the mortality rate among children under five

4.1 Under-five mortality rate
4.2 Infant mortality rate
4.3 Proportion of 1 year-old children immunised against measles

Goal 5: Improve maternal health
Indicators

Target 5a: Reduce by three quarters the maternal mortality ratio

5.1 Maternal mortality ratio
5.2 Proportion of births attended by skilled health personnel

Target 5b: Achieve, by 2015, universal access to reproductive health

5.3 Contraceptive prevalence rate
5.4 Adolescent birth rate
5.5 Antenatal care coverage (at least one visit and at least four visits)
5.6 Unmet need for family planning

Goal 6: Combat HIV/AIDS, malaria and other diseases
Indicators

Target 6a: Halt and begin to reverse the spread of HIV/AIDS

6.1 HIV prevalence among population aged 15–24 years
6.2 Condom use at last high-risk sex
6.3 Proportion of population aged 15–24 years with comprehensive correct knowledge of HIV/AIDS
6.4 Ratio of school attendance of orphans to school attendance of non-orphans aged 10–14 years

Target 6b: Achieve, by 2010, universal access to treatment for HIV/AIDS for all those who need it

6.5 Proportion of population with advanced HIV infection with access to antiretroviral drugs

Target 6c: Halt and begin to reverse the incidence of malaria and other major diseases

6.6 Incidence and death rates associated with malaria
6.7 Proportion of children under 5 sleeping under insecticide-treated bednets
6.8 Proportion of children under 5 with fever who are treated with appropriate anti-malarial drugs
6.9 Incidence, prevalence and death rates associated with tuberculosis
6.10 Proportion of tuberculosis cases detected and cured under directly observed treatment short course

Goal 7: Ensure environmental sustainability
Indicators

Target 7a: Integrate the principles of sustainable development into country policies and programmes; reverse loss of environmental resources

Target 7b: Reduce biodiversity loss, achieving, by 2010, a significant reduction in the rate of loss

7.1 Proportion of land area covered by forest
7.2 CO_2 emissions, total, per capita and per \$1 GDP (PPP)
7.3 Consumption of ozone-depleting substances
7.4 Proportion of fish stocks within safe biological limits
7.5 Proportion of total water resources used
7.6 Proportion of terrestrial and marine areas protected
7.7 Proportion of species threatened with extinction

Target 7c: Reduce by half the proportion of people without sustainable access to safe drinking water

7.8 Proportion of population using an improved drinking water source
7.9 Proportion of population using an improved sanitation facility

Target 7d: Achieve significant improvement in lives of at least 100 million slum dwellers, by 2020

7.10 Proportion of urban population living in slums

Goal 8: Develop a global partnership for development
Indicators
Target 8a: Develop further an open, rule-based, predictable, non-discriminatory trading and financial system Includes a commitment to good governance, development and poverty reduction – both nationally and internationally
Target 8b: Address the special needs of the least developed countries Includes: tariff and quota free access for the least developed countries' exports; enhanced programme of debt relief for heavily indebted poor countries (HIPC) and cancellation of official bilateral debt; and more generous ODA for countries committed to poverty reduction
Target 8c: Address the special needs of landlocked developing countries and small island developing States (through the Programme of Action for the Sustainable Development of Small Island Developing States and the outcome of the twenty-second special session of the General Assembly)
Target 8d: Deal comprehensively with the debt problems of developing countries through national and international measures in order to make debt sustainable in the long term Some of the indicators listed below are monitored separately for the least developed countries (LDCs), Africa, landlocked developing countries and small island developing States. 8.1 Net ODA, total and to the least developed countries, as percentage of OECD/DAC donors' gross nation income 8.2 Proportion of total bilateral, sector-allocable ODA of OECD/DAC donors to basic social services (basic education, primary health care, nutrition, safe water and sanitation) 8.3 Proportion of bilateral official development assistance of OECD/DAC donors that is untied 8.4 ODA received in landlocked developing countries as a proportion of their gross national incomes 8.5 ODA received in small island developing States as a proportion of their gross national incomes *Market access* 8.6 Proportion of total developed country imports (by value and excluding arms) from developing countries and least developed countries, admitted free of duty 8.7 Average tariffs imposed by developed countries on agricultural products and textiles and clothing from developing countries 8.8 Agricultural support estimate for OECD countries as a percentage of their gross domestic product 8.9 Proportion of ODA provided to help build trade capacity *Debt sustainability* 8.10 Total number of countries that have reached their HIPC decision points and number that have reached their HIPC completion points (cumulative) 8.11 Debt relief committed under HIPC and MDRI Initiatives 8.12 Debt service as a percentage of exports of goods and services
Target 8e: In cooperation with pharmaceutical companies, provide access to affordable essential drugs in developing countries 8.13 Proportion of population with access to affordable essential drugs on a sustainable basis
Target 8f: In cooperation with the private sector, make available the benefits of new technologies, especially information and communications 8.14 Telephone lines per 100 population 8.15 Cellular subscribers per 100 population 8.16 Internet users per 100 population

Source: Reproduced with permission from the United Nations Development Programme http://www.undp.org.

Appendix 2: Useful Websites

Academic Journal of the Green Economics Institute: www.inderscience.com/ijge
African Institute of Corporate Citizenship: www.aiccafrica.com
Aspen Institute Business in Society program: www.aspenist.org
Boston College Centre for Corporate Citizenship: www.bc.edu/corporatecitizenship
Brazilian Hunger Campaign: www.fomezero.org
Business and Human Rights Centre: contact@business-humanrights.org
Business Alliance Against Chronic Hunger: www.weforum.org/hunger
Business for Social Responsibility: www.bsr.org
Business in the Community: www.bitc.org.uk
Carbon Disclosure Project: www.cdproject.net
Clinton Global Initiative: www.clintonglobalinitiative.org
Conservation International: www.conservation.org
CSR Europe: www.csreurope.org
Department for Environment Food and Rural Affairs: www.defra.gov.uk
Dow Jones Sustainability Index: www.sustainability-index.com
Environmental Change Institute: www.eci.ox.ac.uk
Ethical Corporation Institute: www.ethicalcorpinstitute.com
Global Alliance for Improved Nutrition: www.micronutrient.org
Global Alliance for Vaccines and Immunization: www.gavialliance.org
Global Alliance for ICT and Development: www.un-gaid.org
Global Business Coalition on HIV/AIDS: www.gbcimpact.org
Global Compact: www.unglobalcompact.org
Global Environmental Management Initiative (GEMI): www.gemi.org
Global Fund to Fight AIDS, TB and Malaria: www.who.int
Global Health Council: www.globalhealth.org
Global Movement for Children: www.gmfc.org
Green Economics Institute: www.greeneconomics.org.uk
Greenhouse Gas Protocol: www.ghgprotocol.org
Initiative for Global Development: www.igdleaders.org
International Alert: www.international-alert.org
International Business Leaders Forum: www.ibif.org
International Chamber of Commerce: www.iccwbo.org
International Institute for Environment and Development: www.iied.org
International Organisation for Employers: www.ioe-emp.org
International Youth Foundation: www.iyfnet.org
Micro-Enterprise Development Programme: www.medep.org
Millennium Ecosystem Assessment: www.millenniumassessment.org
NASA: www.visibleearth.nasa.gov
National Business Initiative: www.nbi.org.za

Optimum Population Trust: www.optimumpopulation.org
Ploughshares: www.ploughshares.ca
Population Resource Centre: www.prcdc.org
The Prince's Rainforests Project: www.princesrainforestsproject.org
The Prince's Youth Business International: www.youth-business.org
RED Campaign: www.joinred.com
Rotary International: www.rotary.org
Roll Back Malaria: www.who.int/rbm
Save the Children: www.savethechildren.net
Stockholm International Peace Research Institute: www.sipri.org
SustainAbility: www.sustainability.com
Transparency International: www.transparency.org
UNEPs Finance Initiative: http://unepfi.net
United Nations International Children's Emergency Fund: www.unicef.org
United Nations Foundation: www.unfoundation.org
United Nations Global Compact: www.unglobalcompact.org
United Nations World Food Programme: www.wfp.org
US Climate Action Partnership: www.us-cap.org
Water and Sanitation for the Urban Poor Initiative: www.wsup.com
Water Aid: www.wateraid.org
WBCSD's Growing Inclusive Markets: www.inclusivebusiness.org
World Economic Forum: www.weforum.org
World Business and Development Awards: www.iccwbo.org
World Business Council for Sustainable Development: www.wbcsd.ch
World Bank: www.worldbank.org
World Environment Center (WEC): www.wec.org
World Health Organization (WHO): www.who.int
World Resources Institute: www.wri.org
World Wide Fund for Nature: www.wwf.org
World Wildlife Fund (WWF): www.wwf.org.uk

Bibliography

Anderson, Bruce N. ed. Ecologue: The Environmental Catalogue and Consumers Guide for a Safe Earth. New York, USA: Prentice Hall Press, 1990.

Barry, John. 'Towards a model green political economy: from ecological modernisation to economic security'. Int. J. Green Economics, Vol. 1, Nos 3/4, 2007.

Benedict, Richard Elliot, et al. Greenhouse Warming: Negotiating a Global Regime. World Resource Institute, 1992.

Brown, Gordon. 'Marshall Plan for the next 50 years'. Washington Post, 17 December, 2001

Brown, Lester R. Plan B 3.0 – Mobilizing to Save Civilization. New York, USA: W. W. Norton & Company Inc, 2008.

Bruges, James. The Big Earth Book. Bristol, UK: Alistair Sawday Publishing Co. Ltd, 2008.

Capra, Frijof. The Turning Point. New York, USA: Bantam, 1982.

Clarke, Robin and Janet King. The Atlas of Water. London, UK: Earthscan, 2004.

Collier, Paul. The Bottom Billion – Why the Poorest Countries are Failing and What Can be Done About it. New York, USA: Oxford University Press, 2007.

Collins, Mark, ed. The Last Rain Forests – A World Conservation Atlas. New York, USA: Oxford University Press, 1990.

Collins, Jim. Good to Great and the Social Sectors. London, UK: Random House, 2006.

Daily, Herman E and John B. Cobb, Jr. For the Common Good: Redirecting the Economy toward Community, the Environment and a Sustainable Future. Boston, USA: Beacon Press, 1989.

Department of Peace Studies, University of Bradford. Peace and Conflict Research at Bradford. Annual Report 2007. Bradford, UK: Duffield Printers, 2008.

Diamond, Jared. Why do Civilisations Collapse? London, UK: Allen Lane, Penguin Books Ltd, 2005.

Diamond, Jared. 'The end of the worlds as we know them'. New York Times, 1 January, 2005.

Doz, Yves and Mikko Kosonen. Fast Strategy. Harlow, UK: Pearson Education Limited, 2008.

Erich, Pica, ed. 'Running on Empty: How Environmentally Harmful Energy Subsidies Siphon Billions for the Taxpayer. A Green Scissors Report'. Washington DC: Friends of the Earth, 2002.

Faulk, Richard A. The Endangered Planet: Prospects and Proposals for Human Survival. New York, USA: Vintage Books, 1971.

Friedman, Thomas L. Hot, Flat and Crowded: Why the World Needs a Green Revolution – and How We Can Renew our Global Future. London, UK: Allen Lane, 2008.

George, Susan. The Debt Boomerang – How the Third World Debt Harms Us All. London, UK: Pluto Press, 1992.

Goodall, Chris. How to Live a Low-Carbon Life. London, UK: Earthscan, 2007.

Gordon, Anita and David Suzuki. It's a Matter of Survival. Cambridge, USA: Harvard University Press, 1991.

Harney, Alexandra. The China Price – The True Cost of Chinese Competitive Advantage. New York, USA: The Penguin Press, 2008.

Henson, Robert. The Rough Guide to Climate Change. London, UK: Rough Guides Ltd, 2008.

Hillman, Mayer. How We Can Save the Planet. London, UK: Penguin Books, 2004.

Houghton, John. Global Warming. Cambridge, UK: Cambridge University Press, 2004.

Ietto-Gillies, Grazia. Transnational Corporations and International Production. London, UK: L Edward Elgar Publishers Ltd., 2005

Ikeda, Daisaku. For the Sake of Peace. Santa Monica, USA: Middleway Press, 2002.

Intergovernmental Panel on Climate Change (IPCC). Climate Change 2007: Impacts, adaptation and vulnerability. Contribution of Working Group II to the Fourth Assessment Report of the IPCC. Cambridge, UK: Cambridge University Press, 2007.

IPCC. Climate Change 2007: Mitigation of Climate Change. Contribution of Working Group III to the Fourth Assessment Report of the IPCC. Cambridge, UK and New York, USA: Cambridge University Press, 2007.

John Paul II. 'The Ecological Crisis a Common Responsibility: Message of his holiness for the celebration of the World Day of Peace'. 1 January, 1990.

Kennet, Miriam and Volker. Heinemann. 'Green economics: Setting the scene', Int. J. Green Economics, Vol. 1, Nos 1/2, 2006.

Kinsley, Michael. Creative Capitalism – Essays by Bill Gates, Warren Buffett and Others. London, UK: Simon & Schuster UK Ltd, 2009.

Kotten, David C. Getting to the 21st Century: Voluntary Action and the Global Agenda: West Hartford, USA: Kuminara Press, 1990.

Krupp, Fred and Miriam Horn. Earth: The Sequel. New York, USA: W. W. Norton & Company Inc, 2008.

Litvinoff, Miles and John Madeley. 50 Reasons to Buy Fair Trade. London, UK: Pluto Press, 2007.

Lomberg, Bjørn. Cool It – The Skeptical Environmentalist's Guide to Global Warming. London, UK: Marshall Cavendish Limited and Cyan Communications Limited, 2007.

Lovelock, James. The Revenge of Gaia. London, UK: Penguin, 2007

Lunn, Claire. E. 'The role of Green Economics in achieving realistic policies and programs for sustainability'. Int. J. Green Economics, Vol. 1, Nos 1/2, 2006.

Mackay, Richard. The Atlas of Endangered Species. London, UK: Earthscan, 2005.

Martin, James. The Meaning of the 21st Century. London, UK: Transworld Publishers, 2006.

McClintock, John. The Uniting of Nations – An Essay on Global Governance. Brussels, Belgium: P.I.E Peter Lang, 2007.

Millstone, Erik and Tim Lang. The Atlas of Food. London, UK: Earthscan, 2008.

Monbiot, George. Heat – How to Stop the Planet from Burning. Cambridge, USA: South End Press, 2007.

'Managing Planet Earth'. Scientific American Special Issue, September 1989.

National Geographic. Special Report: Changing Climate, June 2008.

Palmer, Martin and Victoria Finlay. Faith in Conservation. Washington, DC: The International Bank for Reconstruction and Development/The World Bank, 2003.

Pauling, Linus and Daisaku Ikeda. A Lifelong Quest for Peace. Boston, USA: Jones & Bartlett Publishers. 1992.

Patten, Chris. What Next? Surviving the Twenty-First Century. London, UK: Allen Lane, 2008.

Pisani, Elizabeth. The Wisdom of Whores. London, UK: Granta Publications, 2008.

Riddell, Roger C. Does Foreign Aid Really Work? New York, USA: Oxford University Press, 2007.

Roberts, Paul. The End of Food. London, UK: Bloomsbury Publishing, 2008.

Sachs, Jeffrey. Common Wealth – Economies for a Crowded Planet. London, UK: Penguin Group, 2008.

Sachs, Jeffrey. The End of Poverty. London, UK: Penguin Group, 2005.

Sachs, Jeffrey. 'One tenth of one percent to make the whole world safer' Washington Post. 21 November, 2001.

Sachs, Jeffrey. 'Can extreme poverty be eliminated?' Scientific American, September 2005.

Senge, Peter et al. The Necessary Revolution. London, UK: Nicholas Brealey Publishing, 2008.

Smith, Dan. The Atlas of War and Peace. London, UK: Earthscan, 2003.

Steinberg, Jonny. Three Letter Plague: A Young Man's Journey through a Great Epidemic. London, UK: Vintage Books, 2009.

Stern, Nicholas. A Blueprint for a Safer Planet. London, UK: The Bodley Head, 2009.

Stutely, Richard. The Definitive Business Plan. London, UK: Prentice Hall. 2007.

Suzuki, David and Holly Dressel. Good News for a Change. How Everyday People are Helping The Planet. Vancouver, Canada: Greystone Books, Margate, UK. 2003.

Tudge, Colin. Feeding People is Easy. Italy: Pari Publishing, 2007.

United Nations Development Programme. Human Development Report 2006. Beyond Scarcity: Power, Poverty and the Global Water Crisis. New York, USA: Palgrave Macmillan, 2006.

United Nations Environment Programme. UNEP Year Book 2008. An Overview of Our Changing Environment. Nairobi, Kenya: DEWA and UNEP, 2008.

United Nations Environment Programme. Global Environment Outlook GEO_4 environment for development. Valletta, Malta: Progress Press Ltd, 2007.

UNFPA. State of the World Population 2006. A Passage to Hope. Women and International Migration. New York, USA: United Nations Population Fund, 2006.

UNFPA. State of the World Population 2006. Moving Young. Youth Supplement. New York, USA: United Nations Population Fund, 2006.

United Nations World Population Estimates and Projections. World Populations Prospects. The 2004 Revision. Volume III: Analytical Report. United Nations Publication, 2006.

Ury, William. Getting to Peace – Transforming Conflict at Home, at Work, and in the World. New York, USA: Viking Penguin, 1999.

Walker, Gabrielle and Sir David King. The Hot Topic. How to Tackle Global Warming and Still Keep The Lights On. London, UK: Bloomsbury, 2008.

Williams, Michael. Deforesting the Earth. Chicago, USA: Chicago University Press, 2006.

The World Bank. Atlas of Global Development. Washington DC, USA: The World Bank, 2007.

World Energy Council. Energy Efficiency Policies and Indicators, London, 2001.

The World Watch Institute. State of the World 2008: Innovations for a Sustainable Economy. New York, USA: W W Norton & Company, 2008.

WWF. Living Planet Report 2006. Gland, Switzerland: WWF, 2006.

Young, Zoe. A New Green Order – The World Bank and the Politics of the Global Environment Facility. London, UK: Pluto Press, 2002.

Index

In subheadings, 'GCs' means 'Global Challenges'. Page numbers in *italics* refer to figures and tables.

80:20 Rule 200, 201, 243–44, 262, 293

Abdel-Ahad, Mohamed 35
acting locally 36, 82, 197
action plans 167, 169, 169–70, 183–84, 294, *294*
    2030 Mission 166, *166*
    objectives and strategies
        global 171–81
        individual/organisational *184, 185,* 185–87
    'Yes we can' attitude toward 181–82
Afghanistan 131, 239, 241
Africa
    agriculture 82–83, 84–85, 87
    aid programmes 234, 264, 266
    corruption 272
    deforestation 91
    desertification 92
        reclaiming land 97, 98
    education 121, 123
    fisheries 80
    Green Revolution 264
    health 4, 276
    HIV/AIDS 111–12, 115–16, 117–18, 282–83
    hunger 16, 81, 84–85, 103
    infections disease 112
    marine parks 95
    partnerships 276
    peace agreements 134
    population growth 103, 105
    poverty 102
    reforestation 48
    water supplies 3, 69, 76

agents of change 10, 46, *167*, 167–68, 169, *231*, 231–32
Aggrey, Dr. J.E. Kwegyir 119
aging population 33–34
agriculture 63, 78
    Climate Change's impact on 78–79
    Colony Collapse Disorder (CCD) 94
    energy use 3, 58, 63
    Green Revolution 83–84
    New Green Revolution 84, 285–86
    organic farming 86, 87
    subsidies 106, 141
    topsoil erosion 90, 97
    water use 15, *70*, 70, 71, 75, 77, 78, 81
    Worldwatch Institute's initiatives 82–83
    see also Food Supplies
aid programmes 142, 234, 238–39
    in Afghanistan 241
    in Africa 234, 264, 266
    agricultural programmes 84–85
    best practice global management in 239–40, 242–44
    cooperation/coordination of 266
    education programmes 105, 124
    EU commitments 234, 238
    financial crisis, impact of 103–4, 302–3
    focusing 105–6
    food distribution programmes 105
    health programmes 32, 105, 117
    of NGOs 241
    'problem with the plumbing' 4, 101, *102*, 102–3
        fixing the leaks 106, 242, *243*
    see also philanthropy
air travel 51, 53, 54, 63
alcohol use 111
Algeria 98
Alliance for a Green Revolution in Africa (AGRA) 264

alternative technologies 43, 63, 64
Angola 132
Annan, Kofi 69, 84–85, 248, 270, 272
Antarctic 41
appliances, household 61, 62, *65*, 65–66, 77
aquaculture 81
aquifers 26, *69*, 71, 74, 78, 152
Argentina 97
ARROW approach
    for Climate Change slow-down *51*, 51–52
    for energy sustainability 58, 64–67
attitude, 'Yes we can' 181–82
Australia 46, 75, 77, 79, 97, 124, 239
Austria 76
automobiles 54, 62, 196
aviation 51, 53, 54, 63
awareness, global 5

Bangladesh 79, 81, 86, 115, 124
    micro-financing programme 141, 142, 250
Barker, Joel 159
bees 94
best practice management principles 11, 22
    executive brief examples
        biodiversity 95, 96–99
        climate change 44, 46, 47–48
        conflict avoidance/resolution 133 37
        education 123–25
        energy supplies 60–61, 64
        financing sustainability 140–43
        food supplies 85–86, *85–86*
        health 113–18
        population stabilization 34–37
        poverty eradication 103–4, 105
        water supplies 75–76, 76, 77
    global management 189–90
        customer-driven organisation 193–94, *194*
        effective business planning *191*, 191–93
        effective change management 206–9, *208*
        focus 200
        information technology use 201–5, *202*, *204*
        partnerships and teaming 196–98
        productivity 200–201
        sensitivity to external environment *195*, 195–96
        shared values *205*, 205–6
        understanding competition 198–99
    health check lists 215, 216
        best practice principles *216*, 216
        management information *221–22*
        managing change *223*
        managing innovation *224–25*
        managing motivation *226–27*
        organisation dynamics *219–20*
        organisation structure and roles *217–18*
    international organisations, applied to 213–14, 239–44
    public sector management, applied to 211–13
Bill and Melinda Gates Foundation 22, 114, 266
    programmes sponsored by 82, 116, 122, 250–51, 264, 276
bio-fuels 15, 59, 81–82, 91, 92
Biodiversity
    business plan to address 175–76
    current situation 90–94
    endangered species protection 97
    interconnectivity to other GCs 91, 94, 96, 151, *151–52*
    population growth 89, 90, 92, 93
    opportunities 95–96
    roles
        for businesses 98
        for governments 96–98
        for people 98–99
    species loss 4, 89, 90–91, 93–94, 95
    summary 89–90
    wildlife parks 95, 97
biomass power 61
Body Shop 206
Borlaug, Norman 83, 85
Botswana 117–18, 255–56
Brahma Kumaris 109

Brazil 91, 92, 97, 121, 124, 276
British Telephone (BT) 260, *261*
Brown, Lester 6–7, 135, 146
Buddhism 288
building construction 62–63
Burroughs, William S. 1
business planning 169–70, *191*, 191–93, 282
    for international organisations 239–40, 242, 242–43
business planning framework 7, 10–11, 21–22, 155–56
business plans *170*, 170, 192–93
    external pressures, responding to *195*, 195–96
    global 171–81
    individual/organisational 183–84, *184*, 185–87
businesses
    best practice management principles *see* best practice management principles
    Connected Reporting Framework 260, *260*
    contribution of 248–52, *249*, *271*, *277*
        multinational companies 253–55
        risks and benefits 256–58
    Corporate Social Responsibility (CSR) Audit 258–60
    greenhouse gas emissions 50
    leveraging 247, 278–79
        for anti-corruption management 270–74, *274*
        for conflict management 269–70, *270*
        for corporate social responsibility and sustainability 261–64
        for good governance *249*
        investments to fight GCs 274–75
        multinational companies 252–56
        for philanthropy 264–66
        state owned enterprises 247–48
        for supply chain management 266–69
    oil scarcity, impacts of 59
    partnerships and teaming 275–78
    roles
        as agent of change 168

        for biodiversity 98
        for Climate Change slow-down 50–53, *51*
        for conflict resolution and peace 136
        for education 124
        for energy sustainability 64–65
        for financing sustainability 144–45
        for food supplies *85–86*, 85–86
        for healthcare 117
        for planet sustainability 98
        for population stabilization 36–37
        for poverty eradication 107
        for water supplies 77
    static vs. future-oriented *208*
    sustainability committment 302
    water use 73

California 47, 49–50, 71, 75
Cameroon 71
Canada 81, 97, 239
Cap and Trade system 44, 47–48, 50, 63, 140, 142, 145
carbon dioxide (CO_2)
    absorption by oceans 15
    atmospheric levels of 15, 40
    reducing emissions 43–46, *45*, 98
        business' role in 50–53
        Cap and Trade system 44, 47–48, 50, 63, 140, 142, 145
        government's role in 47–50
        people's role in 53–55
    travel emissions 63, *66*
    see also greenhouse gas emissions
carbon trading 52, 95
cars 54, 62, 196
Chad 71, 265
Challenges faced 5, 25–27
    interconnectivity of *see* interconnectivity
    Perfect Opportunity view 16–17, *154*, 154–55
    Perfect Storm view 14, *14*, 15–16, 153–54
Chambers, Ian 300
change management 169, 206–9, *208*
    for international organisations 240, 244

charitable giving 107, 117, 125, 234, 302–3
Child's Challenge 298
China 16, 43, 195, 237–38, 245–46, 247, 268
  biodiversity loss 93, 94
  deforestation 91
  desert land, reclaiming 98
  education 122
  emissions 150
  energy efficiency 60
  health 112, 115
  reforestation 48
  water supplies 71, 73, 76, 78
Christianity 288
Churchill, Winston 139, 298
civilisation collapse 18–21
clean coal energy 60
Climate Change 3, 25
  adaptation plans 43
  ARROW approach to *51*, 51–52
  best practice examples 46
  business plan to address *172–73*, 183–84, *184*
  cooperation for action 43
  current situation 40–43
  EU commitments 233–34
  greenhouse effect 39
  impacts of *41*, 41–43, *42*, 147
    on agriculture 78–79
    on biodiversity *152*
    on conflict 152–53, *153*
    on food supplies 26, *41*, 78–79
    on health 110, 150, *151*
    on poverty 103
    on security 132, *133*
    on water supplies 15, *41*, 73, 74, 152–53, *153*
  impacts on
    from greenhouse gases 57, 58
    from population growth *15*, 15–16, 40
  key considerations 43–45
  mitigation policies 48, *48–49*
  opportunities 45–46
  partnerships and teaming 197
  preferred maximum temperature increase 44–45
  roles
    for businesses 50–53, *51*
    for governments 47–50, *48–49*
    for people 53–55
  summary 39–40
  tipping points 41
collaboration 47, 165, 197–98, 259
  see also cooperation/coordination
Colony Collapse Disorder (CCD) 94
Common Wealth (Sachs) 146, *147*
communication 36, 46, 52–53, 134, 182, 244, 256
community support programmes 37, 117
competition 196, 198–99, 212
complacency 17, 21, 83
computers 52–53, 122, 123, 202
conceptual frameworks 6–7
Conflict 4, 27
  avoidance of 20, 23, 76
  business plan to address *179–80*
  causes of 16, 20, 132, *133*
    water shortages 9, 15, 73, 152, *153*
  current situation 128–31, *130*
  impacts of 128, *129*
  interconnectivity to other GCs 132, 152–53, *153*, 154
  ongoing in 2007 *130*
  opportunities 133–34
  roles
    for businesses 136, 269–70, *270*
    of governments 134–36
    for people 136–37
  summary 127–28
Congo 131, 132
cooperation/coordination 10, 11, 169, 213, 250, 281
  of aid 106, 266
  for biodiversity 97
  and competition 199
  for defence spending 144
  for financing sustainability 143
  for food supplies 82
  for healthcare 114
  increase of 5
  need for 6, 16, 16–17, 43
  for population stabilization 36

between public and private sectors
    251–52
  success stories 16–17, 144, 213, 303
  Triangle of Change *167*, 168, *231*,
    231–32, 281
  for water supplies 76
  'Yes we can' attitude 181–82
Corporate Social Responsibility (CSR) Audit
    258–60
  issues and best practice examples
    261–64
corruption 136, 189, 237, 241, 242, 254,
    270, 272–74, *274*
cost savings 50, 64
costs, economic
  of achieving global goals *147*
  of addressing social and earth
    restoration goals 145–47, *146*
  of conflict 4, 128, 129, *129*, 131
  of deforestation 92
  of disease 112
  of environmental damage 4
  as investments 148
  see also Financing Sustainability
Coyne, Jerry 89
customer focus 193–94, *194*, 212, 213, 240,
    242, 243

Davies, Mervyn 252
de Beers 255–56
Dead Aid (Moyo) 239
Debswana Diamond Company 255–56
debt cancellation 104, 106
deforestation 19, 46, 48, 91–92
  reducing 95, 96, 98–99
delivering the Plan 291–92, *292*
  being passionate 296–97
  participating *295*, 295–96
  personalising 292–93
  planning *294*, 294–95
  prioritizing *293*, 293–94
Denmark 258
desertification 92, 245
  solutions to 97–98
Diamond, Jared 18
diarrhoeal disease 112, 254
diet, impacts of 34, 78

DiPiazza, Sam 247, 260
disease *see* Health
Donne, John 13
droughts 15, 75, 79
Drucker, Peter 189
drug use 113

Earth *297*
Earth Charter *160*, 160–66
'ease of use' approach 46, 296
East Timor 240
Easter Island case study 18–21
ecosystems 32, *41*, 76, 92
Education 4, 27, 36, 105
  business plan to address *178–79*
  current situation 120–22
  and escape from poverty 105
  impact of *119*
  interconnectivity to other GCs *120*,
    120–21, *121*
  opportunities and best practice 123
  roles
    for businesses 124
    for governments 123–24
    for people 125
  summary 119–20
  of women 10, 34, 35, 37, 114, 119, 121,
    123, 124
Egypt 32, 73, 76
Ehrich, Paul 78
Einstein, Albert 94
The End of Poverty (Sachs) 102–3
energy poverty 59
Energy Supplies
  alternative sources 59, 60–61
  ARROW approach to
    in businesses 64–65
    in homes 66–67
  bio-fuels vs. food production 59, 81–82
  business plan to address *173–74*
  carbon-based vs. alternatives *59*
  current situation 58–59
  efficient use of 61, 61–62, 62–63
  and food production 81–82
  human dependence on 67–68
  interconnectivity to other GCs 81–82
  opportunities 61–62

roles
    for businesses 64–65
    for governments 62–63
    for people 65–68
    summary 57–58
    sustainable 25
energy use 3, 58, *58*, *62*, *65*, 67, 140
    reducing 62–68
Environmental Children's Organisation (ECO) 298
Ethiopia 32, 76, 81, 83, 87, 124
European Union (EU) 63, 80, 159, 233, 234, 238
    climate change commitments 40, 44, 45, 233–34
    water supplies 3, 75, 76, 234
executive briefs 27
    Climate Change 39–55
    conflict and peace 127–37
    education 119–25
    energy supplies 57–68
    financing sustainability 139–48
    health 109–18
    interconnectivity of GCs 149–56
    planet sustainability and biodiversity 89–99
    population growth 29–37
    poverty 101–7
    water and food supplies 69–87
external pressures *195*, 195–96, 200
extreme weather events *41*
eyeglasses, access to 113

Fair Trade 143, 206, 267
faith groups 287–90
family planning 29, 30, 102, 105
    access to 32, 34, 36, 37
    Iranian example 34–36
farming *see* agriculture
fear 17
finances, global 4, 27, 47–48, 140, 143–44
financial crisis, current 301–2
    aid, impact on 103–4, 302–3
    cooperative response to 16, 135, 144, 182, 213, 303
Financing Sustainability 147–48
    business plan for *180–81*

costs, estimating 145–47, *146*, *147*
current situation 140
roles
    for businesses 144–45
    governments 143–44
    for people 145
solutions and best practice 140–43
summary 139–40
fisheries 79–81, *80*, 82, 95
floods 79
Food Supplies 3, 26, 104
    business plan to address *174–75*
    current situation 78–82
    impacts on 16
        from bio-fuel production 59, 81–82
        from Climate Change 26, *41*, 78–79
        from deforestation 92
        from population growth 9, 26, 69, 78, 83–85, 152
    increasing 96, 254
    interconnectivity to other GCs 78, 81, 94
    roles
        for businesses 85–86, *85–86*
        for governments 82–85
        for people 86–87
    summary 69–70
    water supplies, connection to 78
    see also agriculture
forests 4, 46, 48, 91–92
    saving/restoring 95, 96, 98–99
fossil fuels 58–59
France 33, 76, 122, 131
Frances, Juana 39
Franklin, Benjamin 231
Friedman, Thomas L. 57, 63
Fuller, Buckminster 149

Gandhi, Mahatma 132
Geisel, Theodor Seuss 291
geothermal power 61
Germany 11, 63, 98, 131
glaciers 15, 39, 73, 76, 246
global warming *see* Climate Change
Gore, Al 181, 285
governance 84, 134, 193, *249*

governments
  best practice management applications
    to international organisations
      213–14
    to public sector management
      211–13
  Climate Change response 44
  leveraging 244–46
  roles
    as agent of change 168
    for biodiversity 96–98
    for Climate Change slow-down
      47–50, *48–49*
    for conflict resolution and peace
      134–36
    for education 123–24
    for energy sustainability 62–63
    for financing sustainability 143–44
    for food supplies 82–85
    for healthcare 115–16
    for planet sustainability 96–98
    for population stabilization 36
    for poverty eradication 105–6
    for water supplies 76
Greece 239
The Green Imperative, B4E Summit, 2009
  301
Green Revolution 78, 83–84, 264
  New Green Revolution 57, 64, 84,
    144–45, 285–86
greenhouse gas emissions *39*, 40, 58,
    150–51
  of businesses 48, 50
  and deforestation 46, 48
  by forms of travel *66*
  of people *53*, 53–54
  population growth, impact of 32
  reduction of 44, 45, 52–53, 54, 64, 233
  responsibility for action *see* roles:for
    Climate Change slow-down
  urgency of situation 45
  see also carbon dioxide (CO_2)
greenhouse reabsorption 55

habitat destruction 19, 80, 95, 151
Halle, Mark 231
Hardin, Gerret 148

Health 4, 26–27
  business plan to address *177–78*
  current situation 110–13
  inequalities 113
  integrated approach 114
  interconnectivity to other GCs 110,
    111, 120, 121, 150–51, *151*, 154
  opportunities and best practice 113–14
  reproductive 32, 34, 35, 36, 105
  roles
    for businesses 117
    for governments 115–16
    for people 117–18
  success stories 5, 250–51
  summary 109–10
healthcare 34, 37, 102, 106, 109, 113, 116
  reproductive 36, 37, 102, 114
Hinduism 288
HIV/AIDS 4, 84, 111–12, 114, 115–16
  Love Africa Project 282–83
  prevention of 37, 106, 115
  social stigma 117–18
Hoekstra, Hopi E. 89
Hot, Flat and Crowded (Friedman) 63
human progress *13*, 13–14, *286*
Humble, John 258
hunger 81, 84–85, 103
  relieving 104, 105

idealism 22–23, 303
illiteracy 35, 105, 120–21
India 76, 78, 81, 83, 92, 237–38, 247–48,
    265
  corruption 237
  health 112, 115
  influence of people 289
Indonesia 83, 91, 136
Information and Communications
    Technology (ICT) 52–53, 123
  for best practice management 201–5
    data overload *204*, 204
    trust 203, *204*
    waves of development *202*
innovation 14, 63, 64, 87, 145, 197–98, 281
  business leadership in 22, 50, 145, 168,
    247, 253
  managing 224–25

interconnectivity 20, 37, 96
    of agents of change 10, *167*, 167–69, *168*
    of GCs 7, 9, 14, *14*, 84, *149*, 149–53, *150*
        see also specific GCs
    of organisations fighting GCs 153
    Perfect Opportunity view *154*, 154–55
    Perfect Storm view 153–54
    of solutions 9–10, *154*, 154–55
International Monetary Fund (IMF) 233, 236
international organisations 135, 235
    best practice global management in 213–14, 235–42
        increasing effectiveness 242–44, *243*
    leveraging 233–35
    investments 275–76
        in communities 107, 265
        in developing countries 107, 253–54
        in education 123, 124
        in food production 82
        in innovation 145
        in renewable energy 44, 60
        in socially responsible funds 142
        in sustainable technologies 6, 47, 61, 63, 142, 145
Iran 34–36, 124
Islam 288–89
Israel 71, 76, 134

Japan 122, 131, 211
Juran, J.M. 200
"Just Let me Breathe" (Chambers) 299

Karl, Bernie 139
Kenya 99, 134
Ki-moon, Ban 275, 277, 278, 301

lakes 71–72
literacy 35, 105, 120–21

malaria 103, 112, 114, 116, 243, 266
Malawi 85, 91, 243
management, effective *see* best practice management principles
Mandela, Nelson 119
Mead, Margaret 281, 291

Mexico 74, 83, 92, 122, 123, 152
micro-financing 141, 142, 250
migration 32–33
military spending 4, 128, *129*, 129–31
    refocus on GCs 135, *136*, 143, 144
Mongolia 98, 290
Morocco 76, 98
Moyo, Dambisa 239
Mozambique 134
multinational corporations (MNCs) 252–56
mutual assured destruction syndrome (MAD) *127*

Nepal 59, 87, 115
New Green Revolution 57, 64, 84, 144–45, 285–86
New Guinea 91
Niger 71
Nigeria 59, 71, 81, 254–55, 257, 276
Non-Government Organisations (NGOs) 212, 234, 241–42, 250
    criticisms of 235, 241
    partnerships 196, 197, 257, 282
    roles 36–37, 136
Northern Ireland 134
nuclear power 61, 238

Obama, Barak 44, 47, 61, 114, 181, 182, 301
objectives, strategies and action plans 167, 169
    agents of change *167*, 167–68, *168*
    business plan 169–70, *170–81*
oceans 15, 60, 79, *80*, 82, 95
    acidification of 15, 41
    fisheries 79–81, *80*, 82, 95
oil 25, 57, 58–59, 61
    consumption of 3, *58*, 58, *62*
Okonjo-Iweaala, Ngozi 254–55
Oppenheim, Jeremy 148
optimism, cautious 17
overexploitation of resources 18–19, *152*
ozone layer 5, 16, 44

Pakistan 73
Pareto Principle (80:20 Rule) 200, 201, 243–44, 262, 293

Paris Declaration on Aid 240
partnerships and teaming 114, 196–98, 242, 252, 274, 275–78
Peace 4, 27, 132, 137
    business plan for *179–80*
    current situation 128–31, *130*
    opportunities and best practice 133–34
    roles
        for businesses 136
        of governments 134–36
        for people 136–37
    summary 127–28
people
    energy consumption (UK) *65*
    greenhouse gas emissions (UK) *53*
    leveraging power of 281–83
    roles
        as agent of change 168
        for biodiversity 98–99
        for Climate Change slow-down 53–55
        for conflict resolution and peace 136–37
        for education 125
        for energy sustainability 65–68
        for financing sustainability 145
        for food supplies 86–87
        for healthcare 117–18
        for planet sustainability 98–99
        for population stabilization 37
        for poverty eradication 107
        for water supplies 77
Peru 116
pharmaceuticals 91
philanthropic foundations 266
philanthropy 107, 125, 142, 264–66, 302–3
Philippines 32, 250–51
Plan B 3.0 (Brown) 135, 146
Plan for the Planet website 11
Planet Sustainability
    business plan to address *170, 175–76*
    commitments to 302–3
    current situation 90–94
    effective financing 141–43
    key elements for achieving 140
    opportunities and best practice 95–96
    population impacts on 89, 90

roles
    for businesses 98
    for governments 96–98
    for people 98–99
    summary 89–90
planning ahead 20, 47
plants 91
pneumonia 112
polio 5, 116, 250–51
pollution 47, 58, 73–74, 113, *152*, 234
The Population Bomb (Ehrich) 78
Population Growth 3, 19, 25, 29
    best practice management approaches 34–36
    business plan to address *171*
    current situation 30–32
    developed vs. developing countries *31*, 31–32
    education, impact of 10, 34, 35, 37, 121
    EU commitments to stabilization 234
    forecasts *30*, 30–32
    impacts of 20, 32–34
        on Climate Change *15*, 15–16, 40
        on conflict 132
        on food and water supplies 26, 69, 78, 83–85, 152
        on poverty 26, 103
        on sustainability and biodiversity 89, 90, 92, 93
    opportunities 34
    roles for stabilization
        for businesses 36–37
        for governments 36
        for people 37
    summary 29–30
Porter, Michael 279
Portugal 239
Poverty 4, 26, 140
    best practice management approaches 103–4
    business plan to address *176–77*
    current situation 102–3
    interconnectivity to other GCs 26, 103, 105, 132
    opportunities 104–5
    roles for eradication
        for businesses 107

for governments 105–6
for people 107
summary 101–2
wealth distribution *101*
world's 40 poorest countries 105
productivity, improvement in 200–201, 211, 212, 213
public sector management 211–14

rainforests 3–4, 91–92
reforestation 48, 55, 96
renewable energy 23, 47, 54, 57–58, 60–61, 275
research and development 6, 63, 75–76, 122, 196
response, rapid 182
responsibility 10, *167*, 167–68, 183, 232, 292–93, 297
   organisational/individual planning 183–84, 184, *184*, 185, *185*–87
rivers 71, 72–73, 76
road map to a sustainable world 8–9
Roggeveen, Jacob 18
roles *167*, 167–68
   for Climate Change slow-down 47–55, *48–49*, *51*
   for conflict resolution and peace 134–37
   for education 123–25
   for energy sustainability 62–68
   for financing sustainability 143–45
   for food supplies 82–87, *85–86*
   for healthcare 115–18
   for planet sustainability and biodiversity 96–99
   for population stabilization 36–37
   for poverty eradication 105–7
   for water supplies 76–77
Rotary International 5, 116, 250–51
Russia 72, 112, 131, 248

Sachs, Jeffrey 101, 102–3, 146, *147*, 169
Schumacher, E.F. 25
sea levels 39, 40, 79
The Secrets of CEOs (Tappin and Cave) *286*
Seuss, Dr. (Theodor Seuss Geisel) 291
Singapore 75–76
smallpox 116
smart grids 61
Smith, Dan 127
soil erosion 90, 97
solar power 23, 54, 60
solutions
   business plans for
      Climate Change *172–73*
      education *178–79*
      energy supplies *173–74*
      financing sustainability *180–81*
      health *177–78*
      peace *179–80*
      population stabilization *171*
      poverty eradication *176–77*
      sustainability *170*, *175–76*
      water and food supplies *174–75*
   implementation and development 11
   interconnectivity of 9–10, *154*, 154–55
South Africa 94, 95, 134
South Korea 48, 73
Spain 11, 76
species banking 142
spiritual imperative 287–90
state owned enterprises 247–48
Steinberg, Jonny 117–18
Stigson, Bjorn 285
Strong, Maurice 287
subsidies 63, 96, 106, 141
success stories 5–6, 16–17
   acid rain issue 44
   Climate Change slow-down 46, 49–50, 55
   conflict resolution 133–34
   education 122
   energy sustainability 67
   ozone layer issue 44
   polio, fight against 250–51
   population stabilization 35–36
Sudan 32, 76
supply chain management 240, 266–69
sustainable development 142–43
Suzuki, Severn 299
Sweden 142

taxation 142

technology
    alternative 43, 63, 64
    ICT services 52–53
    rapid deployment of 44, 61
Thailand 91, 112
Thinking Globally, Acting Locally principle 36, 82, 196–98, 240, 251
Three Letter Plague (Steinberg) 117–18
tidal power 60
tipping points 41
tobacco use 4, 111, 116, 117
'The Tragedy of the Commons' (Hardin) 148
transport and emissions 53, 54, 63, *66*
Triangle of Change *167*, 168, *231*, 231–32, 281
trust 197, 203, *204*, 205
tuberculosis (TB) 112, 114, 116, 265

Uganda 282
United Kingdom 131, 212–13
    and climate change *53*, 53–54, 55
    energy supplies 60, 62–63
    energy use 65, *65*, 67
    Love Africa Project 282–83
United Nations 213–14, 233, 235–36
    Bruntland Commission Report 143
    Development Programme 251–52
    Global Compact *277*, 277
    Millennium Declaration 114
    Millennium Development Goals 113, 119–20, 167, 305–8
    roles
        for anti-corruption management 106, 273
        for Climate Change slow-down 48, *48–49*
        increasing awareness and coordination 5
        peacekeeping 133, 134–35
        for population stabilization 36
United States 33, 131, 211
    aid contributions 83, 234, 239, 240
    biodiversity 142
    climate change 44, 47, 54
    deserts, reclaiming land from 97
    education 122

    emissions targets 40
    energy consumption *58*
    energy supplies 61, 63
    health 114
    military spending 128, 131
    pollution 113
    water supplies 71, 73, 78
urbanisation 33, 67, 73, 74
urgency of situation 5, 6, 11, 16

values, shared *205*, 205–6
vision statement *160*
    2030 Mission *166*, 166
    Earth Charter 160–66
    foundation principles
        democracy, nonviolence and peace 164–65
        ecological integrity 162–63
        respect and care for the community of life 162
        social and economic justice 163–64
    reasons for 159–60

wars *see* Conflict
water, bottled 73–74
water power 60
Water Supplies 3, 15, 26
    best practice management approaches 75–76
    business plan to address *174–75*
    and conflict 9, 15, 73, 152, *153*
    current situation 70–71
    global usage *70*
    impacts on
        from Climate Change 15, *41*, 73, 74, 152–53, *153*
        from population growth 26, 69, 78, 83–85, 152
    interconnectivity to other GCs 74, 78, 81, 115
    key impacts 71–74
    opportunities 75
    roles
        for businesses 77
        for governments 76
        for people 77
    summary 69–70

weather patterns, changing 15, 39, 40, 43, 74
websites 11, 309–10
wetlands banking 142
Why do Civilisations Collapse? (Diamond) 18
wildlife *see* Biodiversity
Wilson, Edward O. 1, 91, 95
wind power 60, 66
women 152–53
    death in childbirth 110
    education of 10, 34, 35, 37, 114, 119, 121, 123, 124
    family planning education/access 32, 35
World Bank 122, 123, 233, 236–37, 275
World Trade Organisation (WTO) 233, 237–38
Worldwatch Institute 82–83

'Yes we can' attitude 181–82
Yunus, Muhammad 250

Zedillo, Ernesto 251

The Green Economics Institute

The Significance of Green Economics

Green Economics is one of the fastest growing global movements for change. It has been taken up by many governments and NGOs and is influencing the context within which businesses operate. It seeks to reform the very concept of economics itself and to incorporate a new set of factors into business and decision making.

Against this background, businesses of all kinds need to adapt to these changes which provide as many opportunities as they do constraints. Ban Ki Moon, General Secretary of the United Nations, has called this the Age of Global Transformation, the Age of Green Economics.

The Green Economics Institute has spread this message over the last 10 years and is leading thought in this new global transformation. It has rewritten traditional economics and business into an overall more viable and meaningful language. This is designed to include all people everywhere; consider other species, nature, the planet and its systems. Its strategies are participatory, cooperative and inclusive.

The Director of the Institute, Miriam Kennet, is a member of University of Oxford's Environmental Change Institute and the Intergovernmental Panel on Climate Change, IPCC. The Institute has its own delegation to the COP Process and the Kyoto Protocol. Its team includes several Nobel Prize winners, the creator of the Carbon Market and the Kyoto Protocol economics, as well as the Kyoto 2 concept.

The Green Economics Institute runs regular events to help you make sense of this new business landscape and to provide you with strategies for your own businesses. With regular events, offices and activities in Oxford University and Brussels, as well as strong connections with Brazil and China, the Institute is at the forefront of this global revolution. Its ideas are driving the revolution forward and are being used as a leadership model by both the Dow Jones and Wall Street.

Miriam Kennet is Director of the Green Economics Institute and Editor of the first double blind peer reviewed academic journal for Inderscience academic publishers – The

International Journal of Green Economics. She is also on the IPCC Intergovernmental Panel on Climate Change and leads a delegation in the Kyoto Protocol Conferences COP 15 and COP16. She is author of many articles and teaches and lectures all around the world. She led global hi tech projects for a very large telecoms provider and she modeled the new innovations in Green Economics on a combination of the Arctic Monkeys and on France Telecom/Equant's business models.

The Green Economics Institute, 6 Strachey Close, Tidmarsh, Reading, RG8 8EP. United Kingdom email:greeneconomicsinstitute@yahoo.com; Telephone: + 44 118 9 841 026; www.greeneconomics.org.uk

If you have found this book useful you may be interested in other titles from Gower

The Future of Innovation
Eds Bettina von Stamm and Anna Trifilova
528 pages; Paperback; 978-0-566-09213-8

Wealth, Welfare and the Global Free Market: A Social Audit of Capitalist Economics
Ozer Ertuna
244 pages; Hardback; 978-0-566-08905-3

Spirituality and Corporate Social Responsibility: Interpenetrating Worlds
Ed. David Bubna-Litic
244 pages; Hardback; 978-0-7546-4763-8

The Global Business Handbook: The Eight Dimensions of International Management
Eds David Newlands and Mark J. Hooper
612 pages; Hardback; 978-0-566-08747-9

Commoditization and the Strategic Response
Andrew Holmes
244 pages; Hardback; 978-0-566-08743-1

For more titles on Corporate Strategy, please visit:
www.gowerpublishing.com/corporatestrategy

Visit **www.gowerpublishing.com** and

- search the entire catalogue of Gower books in print
- order titles online at 10% discount
- take advantage of special offers
- sign up for our monthly e-mail update service
- download free sample chapters from all recent titles
- download or order our catalogue